# TOPICS IN THE GENERAL THEORY OF STRUCTURES

**THEORY AND DECISION LIBRARY**

General Editors: W. Leinfellner and G. Eberlein

Series A: Philosophy and Methodology of the Social Sciences
Editors: W. Leinfellner (Technical University of Vienna)
G. Eberlein (Technical University of Munich)

Series B: Mathematical and Statistical Methods
Editor: H. Skala (University of Paderborn)

Series C: Game Theory, Mathematical Programming and Mathematical Economics
Editor: S. Tijs (University of Nijmegen)

Series D: System Theory, Knowledge Engineering and Problem Solving
Editor: W. Janko (University of Vienna)

---

SERIES D: **SYSTEM THEORY, KNOWLEDGE ENGINEERING AND PROBLEM SOLVING**

Editor: W. Janko (Vienna)

**Editorial Board**

G. Feichtinger (Vienna), H. T. Nguyen (Las Cruces), N. B. Nicolau (Palma de Mallorca), O. Opitz (Augsburg), H. J. Skala (Paderborn), M. Sugeno (Yokohama).

**Scope**

This series focuses on the design and description of organisations and systems with application to the social sciences. Formal treatment of the subjects is encouraged. Systems theory, information systems, system analysis, interrelated structures, program systems and expert systems are considered to be a theme within the series. The fundamental basics of such concepts including computational and algorithmic aspects and the investigation of the empirical behaviour of systems and organisations will be an essential part of this library. The study of problems related to the interface of systems and organisations to their environment is supported. Interdisciplinary considerations are welcome. The publication of recent and original results will be favoured.

# TOPICS IN THE GENERAL THEORY OF STRUCTURES

*Edited by*

E. R. CAIANIELLO

*Dipartimento di Fisica Teorica, Università di Salerno, Italy*

and

M. A. AIZERMAN

*Institute of Control Sciences, Moscow, U.S.S.R.*

D. REIDEL PUBLISHING COMPANY

A MEMBER OF THE KLUWER  ACADEMIC PUBLISHERS GROUP

DORDRECHT / BOSTON / LANCASTER / TOKYO

**Library of Congress Cataloging in Publication Data**

Topics in the general theory of structures.

(Theory and decision library. Series D, System theory, knowledge engineering, and problem solving)
Includes index.
1. System analysis. 2. Computational complexity. I. Caianiello, Eduardo R., 1921– . II. Aĭzerman, M. A. (Mark Aronovich), 1913– . III. Series.
QA402.T65 1987 003 87–4329
ISBN 90–277–2451–2

---

Published by D. Reidel Publishing Company,
P.O. Box 17, 3300 AA Dordrecht, Holland.

Sold and distributed in the U.S.A. and Canada
by Kluwer Academic Publishers,
101 Philip Drive, Assinippi Park, Norwell, MA 02061, U.S.A.

In all other countries, sold and distributed
by Kluwer Academic Publishers Group,
P.O. Box 322, 3300 AH Dordrecht, Holland.

All Rights Reserved
© 1987 by D. Reidel Publishing Company, Dordrecht, Holland
No part of the material protected by this copyright notice may be reproduced or utilized in any form or by any means, electronic or mechanical including photocopying, recording or by any information storage and retrieval system, without written permission from the copyright owner

Printed in The Netherlands

## ACKNOWLEDGMENT

The Editors express their warm thanks to the Istituto Italiano per gli Studi Filosofici and to the International Institute for Advanced Scientific Studies, Naples, for generously promoting and organizing a series of Meetings in Naples, Capri and Amalfi, from which the collaboration between Italian and Soviet scientists, the object of this volume, has greatly benefited.

TABLE OF CONTENTS

# INTRODUCTION

This volume is about "Structure".

The search for "structure", always the pursuit of sciences within their specific areas and perspectives, is witnessing these days a dramatic revolution.

The coexistence and interaction of so many structures (atoms, humans, cosmos and all that there is in between) would be unconceivable according to many experts, if there were not, behind it all, some general organizational principles that (at least in some asymptotic way) make possible so many equilibria among species and natural objects, fantastically tuned to an extremely high degree of precision.

The evidence accumulates to an increasingly impressive degree; a concrete example comes from physics, whose constant aim always was and is that of searching for "ultimate laws", out of which everything should follow, from quarks to the cosmos. Our notions and philosophy have undergone major revolutions, whenever the "unthinkable" has been changed by its wonderful endeavours into "fact". Well, it is just from physics that evidence comes: even if the "ultimate" could be reached, it would not in any way be a terminal point. When "complexity" comes into the game, entirely new notions have to be invented; they all have to do with "structure", though this time in a much wider sense than would have been understood a decade or so ago. Clear indications show the difference: the laws that determine "structures" in physics (in this volume we shall quote some examples of them) are not of purely physical nature, such as Newton's or Maxwell's laws: they are rather logical, connected with information, entropy, the structuring of complex systems into levels. Such laws, which are by some claimed to hold from hadron jets to galaxy clusterings, do not belong to mechanics, whether classic or quantum. This view is enforced by the fact that exactly the same laws emerge from a variety of fields which have nothing to do with physics: to name one, mathematical linguistics.

The challenge in the study of structure is epitomized by the word "complexity". We have to learn "simple" ways to catch what is really essential in phenomena which occur only in systems with large numbers of components and interactions.



*E. R. Caianiello and M. A. Aizerman (eds.), Topics in the General Theory of Structures, 1–3*
*© 1987 by D. Reidel Publishing Company.*

To give an example from a different field, suppose a situation in which a large number of voters will determine, with their free vote, the election of one among a number of competing candidates; such a situation cannot be reduced to the familiar methodology of physics, of descending to ultimate constituents and ascending back to macroscopic objects. What happens is something that has to be studied by considering the system in its actual complexity; models that will render its handling amenable to mathematical treatment must not destroy that complexity, but only expurge from it all finer details of the real structure that are not relevant in the global perspective.

The works collected in the present volume all point toward this goal. About a decade ago the emerging importance of this subject was recognized in an official agreement between the Italian C.N.R. and the Soviet Academy of Sciences. Two teams were to cooperate in this widely interdisciplinary effort, led respectively by the two authors of this volume. Each year there were meetings and general discussions, from which it soon emerged that the important thing to attempt was not to try to build a general theory of things of which basic features are still unknown, but rather to assault the fortress by a number of well defined inroads, aiming at some concrete, though modest, knowledge which might provide a basis for further speculation in a quest that no sane person can expect to see achieved within his lifetime.

The task is far from finished, the cooperation continues; enough material has been gathered however, to make desirable its publication as a set of contributions, each addressing an aspect of the problem up to a degree of completion permitting tests and applications.

The contributions cover the following aspects of general structure theory.

## 1. HIERARCHICAL SYSTEMS

All complex systems are structured into levels, each containing elements characterized by a value which depends upon the level; "value", ignored in information theory, appears to be essential to understand the structure of systems. As the simplest model, we can consider a counting (decimal or binary) system or a monetary system, which are characterized by their modular structure (e.g. a power of ten, or two). A general theory of such modular hierarchical systems is made, which yields results in surprising agreement with many more facts than one could a priori expect. Urban population, army hierarchies, the Zipf law of linguistics etc. follow straightforwardly. Two ways of change are possible for such systems, "evolution" and "revolution" (change of module); equilibrium between two such systems is possible only if they have the same structure.

## 2. GRAPHODYNAMICS

The structure of systems can in most cases of interest be described by means of oriented graphs (such as armies, corporation structures, computer memories etc.). These graphs contain a relevant amount of information about the system. It therefore becomes of interest to have simple mathematical techniques which may describe with a minimum of formal complication the change of structure through the change of the corresponding graphs.

## 3. STRUCTURAL PROPERTIES OF VOTING SYSTEMS

When the components of a system can express their individual opinion with a decision which will select one out of a set of choices (projects, nominations, plans), many subtle questions of logic and paradoxical situations are encountered; "optimization" may not be possible or even desirable. A survey and study of this situation, which is of the highest interest to social scientists, is made in chapter 4 and in appendix A.

## 4. C-CALCULUS

Complex systems, for instance a written text, a set of images, an assembly of biological cells, are the object of a huge variety of *ad hoc* mathematical studies and techniques. A simple calculus, simpler in fact than arithmetics because it uses only the direct operations of sum and product, is proposed and tested in various such cases with satisfactory results, as a "natural" language of rather general applicability in many such situations. To this "calculus" are dedicated chapters 5,6 and 7.

## 5. UNCERTAINTY IN MODEL BUILDING

It has always been assumed that uncertainty is typical of quantum physics. It has been proved that this is not true; the famous Cramér-Rao inequality of statistics has the same content as Heisenberg's uncertainty relation, and a geometrical approach is open which can work in two ways: quantum mechanics can be derived from it, as shown here; conversely, the well proven methods of quantum mechanics become conceivable in many other systems in which uncertainty plays a role.

## STRUCTURE AND MODULARITY IN SELF-ORGANIZING COMPLEX SYSTEMS

E.R.Caianiello[1] M.Marinaro[1] G.Scarpetta[1,2] G.Simoncelli[3]

(1) Dipartimento di Fisica Teorica e sue Metodologie per le Scienze Applicate - Università di Salerno - Salerno - Italy
(2) Istituto Nazionale di Fisica Nucleare - Sezione di Napoli - Italy
(3) Dipartimento di Scienze di Base - Accademia Aeronautica - Pozzuoli

### 1. SYSTEM AND PATTERN

The mathematical notion of set is a primitive concept, i.e. it cannot be defined in terms of other, simpler concepts. If we say that a set is a collection, or an aggregate, all we are actually doing is just giving its synonyms. The concept of set, in other words, is as primitive as the concept of element, even though the two logical concepts are obviously quite different and indeed even, to some extent, antithetical; one and the some object can be a set and at the same time can be considered an element of a larger set.

More or less the some applies to the term system. Etymologically, system is the Greek equivalent of the Latin composition; it therefore implies, first of all, the simultaneous presence of more than one component, part or organ. A system however is not merely a set of independent elments. The word implies an interaction of the parts such that the totality of the parts presents characteristics and properties which are not the result of a simple addition or juxtaposition.

The following example helps to clarify the concept. Let us take a set of stones (hewn stones) normally used in the building of arches. With the aid of an arch centre as support, the hewn stones are placed one next to the other. Once a keystone (i.e. the central hewn stone) is inserted, the set of stones acquires, thanks to the elements' capacity of supporting one another, the static capacity of supporting not only itself, but even superimposed loads: a structure is born. The load supporting capacity is not a property of each stone, or even of all the stones put together, it comes into being the moment the stones interact in a given arrangement.

Broadly speaking, a system may have dynamic properties, i.e. properties which entail, in time, a change not only in the interactions of the parts, but also in the components themselves. In this sense, living organisms, as well as social, economic or political organizations, are



E. R. Caianiello and M. A. Aizerman (eds.), Topics in the General Theory of Structures, 5–57.
© 1987 by D. Reidel Publishing Company.

all systems.

A system may be _open_ or _closed_. We shall say that a system is open when there is a fundamental interaction not only of the parts, one with the other, but also of the parts and the external environment. A living cell, for instance, is an open system, because in addition to the chemical interactions of its different components, the cell receives nourishment and eliminates waste matter, i.e. there is a continuous exchange with the environment. The personnel of a company, organized according to the specific task each employee is qualified to perform, may be regarded, in many ways, as a closed system (not considering the substitutions which take place from time to time); in fact an exchange with the environment(in this case, the social community within which the firm operates) is not vital to the function of the system. In town-planning, an entire city may be viewed as a system, with the continuous changes in its population (immigrants, emigrants, births, deaths), its industrial development and the transfer of incomes (rise of one social class, decline of another). Languages too may be regarded as systems, each of which is characterized - as a structure - by its own particular morphemes and syntagms, its grammar and its syntax.

In our opinion also the concept of _pattern_ (or, better still the Italian equivalent, _forma_, which gives an even clearer idea of the concept) is primitive. When a common property is found in a number of configurations , we can say that such configurations have the same pattern: the common property is abstracted and regarded as a new mathematical entity, the _pattern_ common to all those configurations which possess it. This notion of pattern can no doubt be related to that of the Platonic idea: but what we are particularly interested in is the fact that the abstraction, _depends, and decisively so, on the person_ who does the abstracting and that the abstraction is quite often _associated to a specific purpose_.

The analysis of patterns requires that the _structure_ be brought out, so that the patterns might be described in terms of other component patterns. An essential feature of this kind of analysis is that it calls for the formation of hierarchies and the description for each level of the characteristics of the patterns of that level.

As a rule, difficulties arise from the fact that we in fact know nothing at all about levels, nor about their characteristics; consequently, the problem of pattern analysis breaks down into a myriad of special sub-problems, a few of which are of a technical nature and many others, viceversa, are of a rather deep and so far unknown nature. The crucial task is no longer to find out whether a given characteristic is present or not, but to discover which _characteristics_ and _how many_ levels there must be, _for the intent we have in mind_.

## 2. QUANTIFICATION AND STABILITY

For many centuries it was held that the colours of the rainbow were seven (just as seven are the musical notes, the cardinal and theological virtues and the deadly sins), but physical analysis has shown that in fact the colours make up a spectrum in which each hue (each wavelength) gradually runs into the successive one (the same of course applies to sounds and, as likely as not, to virtues and sins). Thus, two realities, of a different order, seem to exist: the first, which belongs to the order of nature, is the reality which inspires the physicist's attitude and in which the colours form a continuum; the second, which belongs to the category of the intellect since it was invented by man, is the reality in which colours (as also sounds) are recognized and classified according to a discrete sequence.

We shall call quantification that process by which a continuum is reduced to a discrete. This reduction is of paramount importance in the study of biological and social structures and, more generally, in all those studies whose major concern is man: for the Chinese and the Japanese, as we know, colours, notes (and virtues) are in all five. It is evident therefore that quantification applied by different societies does not lead to the same results.

A comparison with civilizations founded on the use of ideograms brings to light another important, and at first glace quite unique, phenomenon: ideograms, which for the most part were originally pictograms, i.e. drawings which suggested a concept by means of a sketched image, are today taught to Chinese and Japanese pupils without any reference to their original meaning. The spontaneous, primordial process of pictographic writing has thus undergone such a transformation in the course of those civilizations' evolution as to become nothing but a mass of several thousand standard, i.e. quantified, symbols: and this has occurred in spite of the fact that the loss of instant recognition which the original symbols afforded has made those languages infinitely more difficult to learn.

In the light of this, we might well ask: how can such a quantification be justified? The answer lies in the fact that it generated a marvellously efficient system of communications which for thousands of years has bound in one immense common civilization the peoples of China who speak different languages and who often can communicate with each other only by means of these written characters which have undergone no modifications throughout the ages (except for a few recent and much-discussed efforts aimed at a simplification).

In our part of the world also, various phenomena have occurred, phenomena which are different but just as important for the development of society: our language too is subject to quantification. A knowledge of these collective mechanisms which lie at the very roots of an institution, such as writing, which is so fundamental to the growth of a culture, is essential and represents a first, important step towards understanding in what way a civilization, or more generally any social institution, develops.

Let us consider the alphabet: would we be able to read a manuscript had we not learnt an alphabet? The answer, quite obviously, is no: an unfamiliar writing often creates serious problems of interpretation. These problems are overcome only if we manage to decode the rules which permit the signs traced by a hand to be converted into the twenty-one letters of the Italian alphabet (or the twenty-six of English, or the thirty-two of Russian...).

Actually, we would not be able to understand one another even when speaking, had we not in common a scheme of quantified phonemes (a few dozen in all), within which to arrange the continuum of sounds, which each of us utters in a different manner, perhaps with a distorted voice, or in a whisper, or singing. Phonemes, which linguists discovered not too long ago [1], differ from language to language (or from dialect to dialect); a Japanese, for instance, cannot distinguish between our "l"s and our "r"s whereas to our ears many sibilant sounds or the different tones of the Chinese language are quite indistinguishable.

It is interesting to compare the systems of writing based on ideograms with those based on the alphabet. Whereas an ideogram, once learnt, recalls instantly to mind one or often several self-contained concepts, a letter of the alphabet, taken singly, has no meaning at all: in other words, it is an abstract symbol which needs to be linked to other symbols in order to generate a meaning. To this fact, which is peculiar to Indo-European languages, is ascribed the development of certain modes of thinking typical of the western hemisphere and, in general, that kind of analytical rationality which is the fundamental framework of our mind (which, we would like to emphasize, is by no means the only framework suited to describing the world or to carrying out scientific work).

This process of quantification encompasses every field, even the social field: as soon as a community begins organizing itself, the roles and tasks of each one of its members are quantified. These "experimental" facts lead us to one and the same conclusion: that any and every system always presents quantified patterns and structures.

An obvious advantage of quantifying the continuum into the discrete is that this guarantees the stability of the structures thus built. The criterion of stability with respect to fluctuations is a fundamental principle by means of which the human brain organizes information, discretizing and quantifying it in such a manner that, when a disturbance occurs, even a rather serious one, no confusion is created.

This relation between quantification and stability is in our opinion fundamental in any quantitative discussion on structures.

In time a structured system may undergo changes. In this connection a change may belong to one of two different categories: one involves a change in the very structure of the system, the other concerns a modification in the type or number of elements which make up the system, without however altering the kind of interaction or relation between the elements which characterize the structure of the system. We shall call the changes belonging to the first type "revolutions", those belonging to the second type "evolutions".

To illustrate the difference between the two types of change, let us consider a hierarchical system which is isomorphic to a set of numbers expressed in a given base: an evolution of the system entails a change in some of these numbers, its revolution calls for a change in the base of the numerical system, in order to maintain the isomorphism.

Let us here emphasize that we use these terms merely in a technical sense. A literal interpretation, though apparently attractive, could be quite misleading.

## 3. SELF-ORGANIZING HIERARCHICAL SYSTEMS

The above remarks have brought out the important role that structural analysis plays in a cybernetic study of systems. As we have seen, structural analysis consists in finding and recognizing the different hierarchical levels and the characteristics of each level.

On the other hand, if we wish to develop an ad hoc methodology which employs the methods of physics and the techniques of mathematics we must keep within certain limits, that is, we can deal only with those particular systems which present a structure capable of change within a given class of equivalence. We call self-organizing those systems · which are capable of modifying their strucure in order to meet the requirements, satisfy the conditions imposed by the environment with which they interact. We shall limit ourselves to the simplest case possible, the case of self-organizing systems which are also modular [2](such as the decimal numerical system), but which can describe a large number of intersecting systems (in economics [3], linguistics, social sciences [4]).

Let us consider a system made up of N elements distributed on L+1 levels. The elements which belong to the same level cannot be distinguished from those belonging to other levels. If $n_h$ is the number of identical elements belonging to level h (h = 0, 1,...L), the number $W_h$ of possible different states (*) that we can form with them is obviously

$$W_h = n_h + 1$$

This is the method of counting the states in Bose-Einstein statistics. When dealing with non-identical elements, viceversa, we shall apply the count of Boltzmann statistics.

If $n_h$ is the number of elements belonging to level h, the total number of elements is

$$N = \sum_{h=0}^{L} n_h$$

This distribution will be called a <u>partition</u> of the elements of the system.

We shall now introduce the notion of <u>value of an element</u> belonging to level h. This new concept is particularly useful: in fact it will permit us to derive a law on the distribution of the elements at each level and to build a thermodynamics of the states of equilibrium of a hierarchical modular system. Let us assume therefore that a value (integer number) $v_h$ is assigned to each level h, with the aid of some criteria of plausibility; the total value of the system is

$$V = \sum_{h=0}^{L} n_h v_h$$

(*) Two states differ if they comprise different numbers of elements.

the mean value of an element of the system is

$$\langle v\rangle = \frac{\sum_{h=0}^{L} n_h v_h}{\sum_{h=0}^{L} n_h}$$

We require moreover that the value function be such that the ratio between $v_h$ and $v_{h-1}$ be an integer greater than one, for each h, and that further, in convenient units, $v_0 = 1$. If, in particular, we have

$$\frac{v_h}{v_{h-1}} = \frac{v_{h-1}}{v_{h-2}} = \ldots = \frac{v_1}{v_0} = M$$

then the hierarchical system is modular and M is the module or base.

We have already defined as self-organizing systems those systems capable of self-adjustment in relation to the interaction with the external environment, with the "universe", by selecting an appropriate partition of the elements. It is now convenient to introduce a relation which links different partitions of one and the same self-organizing hierarchical system.

Let us say that a partition $\pi_b$ is a refinement of a partition $\pi_a$ if it has more levels of $\pi_a$ and maintains all the levels of $\pi_a$. Consequently, a partition $\pi_b$ of a hierarchical modular system is a refinement of a previous partition $\pi_a$ if, and only if

$$M_b = M_a^{\frac{1}{p}}$$

with p an integer greater than one. This process is a refinement of order p of the system. Let us consider, for instance, a hierarchical system made up of only two levels

$\ell = 1$ ———————— $v_1 = 16$
$\ell = 0$ ———————— $v_0 = 1$

the module is 16. Let us perform a refining transformation of order four:

$$M \longrightarrow M' = \sqrt[4]{16} = 2$$

the levels now are:

$\ell = 4$ ———————— $v_4 = 16$
$\ell = 3$ ———————— $v_3 = 8$
$\ell = 2$ ———————— $v_2 = 4$
$\ell = 1$ ———————— $v_1 = 2$
$\ell = 0$ ———————— $v_0 = 1$

## 4. INFORMATION AND FORMATION OF THE LEVELS

A system is structured so as to present hierarchical "levels", which characterize the role and the "value" of the elements of the system. We can find out about the part which the existence of different levels plays in a system, by analyzing, for instance, the reasons why even rather simple systems, such as monetary systems, are always formed by discrete elements, with a different denomination and value, rather than by a continuum, or by identical elements (apart from the case, obviously, of the monetary or bartering system of primitive societies).

Let us therefore suppose that we have, to begin with, a certain number N of identical coins, all of the same value v. With such coins we shall be able to make payments relative to N+1 distinct values (that is, of value Nv, (N - 1)v and so forth down to v, obviously in addition to the trivial case of no exchange, to which corresponds value zero). The N identical coins can, in other words, be placed in $W = N + 1$ different "states" (all - a priori - equally probable if we have no information on what to purchase).

The information associated to N coins is measured by the quantity

$$I = k \ln W$$

The logarithmic growth of I (almost linear for a small N, but very slow for high values of N) brings out that, as N increases, the system becomes more difficult to handle, no matter what we intend to do with it or find out about it [5] : while for a small N we need only double N to double the information, when N is large, in order to obtain double information, the value of N has to be increased out of all proportion. Let us therefore suppose that, at this point, the system is restructured and that in the new system, instead of bringing in new coins identical to the previous ones, we introduce "clusters" of M coins: each cluster,or, if you prefer, a coin of a different type, will thus represent a new element in the system, with a value Mv, an element which belongs to a new level.

Consequently, the information associated to N coins ($N_1 = M - 1$ of the first type, with value v and $N_2 = N - N_1$ of the second type, with value Mv) will be now

$$I = k \ln(N_1 + 1) + k \ln(N_2 + 1) > k \ln(N + 1)$$

because the different "states" in which N coins can now be distributed are

$$W = (N_1 + 1)(N_2 + 1)$$

Fig.1 illustrates the information function $I = k \ln W$ as N varies.

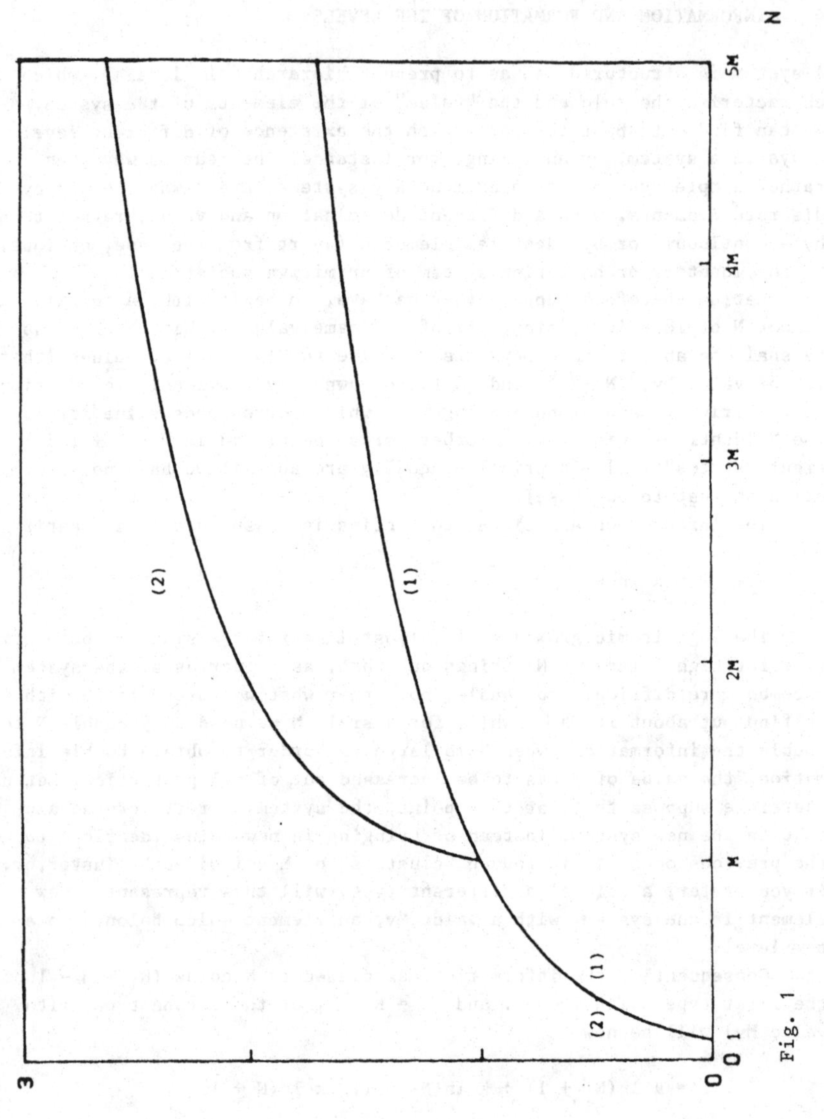


Fig. 1

Curve (1) represents the case where N coins are identical, i.e. they have the same value v. Curve (2) illustrates the case in which the monetary system has been restructured with the introduction of a new level, to which coins with value Mv belong. The figure shows clearly how the information associated to N coins grows almost linearly with N for a larger interval of values and this is a direct result of the introduction of a new level. This interval will become even larger if we introduce a third level with coins of $M^2v$, and so on. Curve (2) in fact can also be viewed as deriving from a contraction, with a scalar factor 1/M of the underlying portion of the horizontal axis; if we group together the coins of Mv value into clusters of M coins (which we consider as coins of value $M^2v$ ), we shall get a further contraction of the horizontal axis for values of N greater than 2M.

What applies to a monetary system obviously holds for any system formed by identical elements, from the point of view of associated information: associated information grows rapidly if the system is reorganized into levels, so that to each level belong clusters of elements of the lower level (each cluster, or group, considered as a single element). It is noteworthy that an element, in becoming part of an element of a higher level, loses many of its own characteristics and acquires the new function of group member.

## 5. HIERARCHICAL MODULAR SYSTEMS [2]

It is important to make a clear-cut distinction between two orders of questions: a) those pertaining to specific assignments of values to the levels of a Hierarchical Modular System (to this order of questions we shall answer in specific examples); b) those which relate to the distribution $n_h$ of the N elements of the system among its levels, h = 0,1,2, ...,L, once their number and values are known; this distribution will reflect the way the system adapts to an external requirement.

We expect that general properties may be derived from a study of the second question.

As long as a self-organizing HS stays "isolated", it has no reason (from what has been said thus far) to prefer any particular choice of $n_h$. If, however, it is in "interaction with the universe" this can no longer be expected to be the case. We shall adopt for handling this problem the following principle: to assume that the system, which is in a given situation, is however at liberty to change itself by any refinement of its original hierarchical partition. This principle assumes, in other words, that the "universe" is not interested into what partition the system choses to organize itself; its interaction with the system is of a "global" nature, to be expressed therefore by requiring that the value of some "mean" quantity of the system be imposed externally.

This principle requires that the average value of an element of an HMS stays invariant under any p-refinement of the HMS

$M \longrightarrow M^{1/p}$ then

$L \longrightarrow pL$

and it is demanded that ($n_h \equiv n_h^{(1)}$ ):

$$(1) \qquad \langle v \rangle = \frac{\sum_{h=0}^{L} n_h^{(1)} M^h}{\sum_{h=0}^{L} n_h^{(1)}} = \frac{\sum_{h=0}^{pL} n_h^{(p)} M^{h/p}}{\sum_{h=0}^{pL} n_h^{(p)}}$$

We also note that an HMS possesses a natural invariant under modular changes: the ratio of the maximum number of states to the module:

$$w = \frac{M^{L+1}}{M} = \frac{(M^{1/p})^{pL+1}}{M^{1/p}} = M^L$$

We must determine $n_h^{(p)}$ so that, for given L,M, $\langle v \rangle$ stays the same for any p. This task, otherwise formidable, turns out to be surprisingly simple for modular systems. In the following section we shall show that the only solution to (1) is the distribution function

$$(2) \qquad n_h^{(p)} = N \frac{1 - M^{-1/2p}}{1 - M^{-1/2p-L/p}} \cdot M^{-h/2p} = n_0^{(p)} M^{-h/2p}$$

The relation (2) is noteworthy: it shows how N elements must, to satisfy the principle of invariance, be distributed among the various levels in relation to each p-refinement. In the following we need not consider values of p other than 1; in an HMS there is no way of knowing a "past history" of levels. In particular we have for p = 1

$$(3) \qquad n_h = n_0 M^{-h/2} = \frac{n_{h-1}}{\sqrt{M}}$$

The number of elements belonging to the level h equals the number of elements belonging to the level immediately below divided by the square root of the module.

Once the law of distribution is found, we may calculate the average value $\langle v \rangle$; we have

$$\langle v \rangle = \frac{1}{N} \sum_{h=0}^{pL} M^{h/2p} = M^{L/2}$$

which is, as expected, independent of p.

We may now try to give a general formulation to the method which has been applied here in extenso only to sub-class (N = const) of HMS's. Asking no questions as to "why", but only as to "how", we take as initial data a description of the levels, and of the value functions attached to them (these define the "structure" of the system). In so doing, one is naturally led to consider transformations, such as regards its interactions with, i.e. response to, its "universe" (as measured by average values) and the type of its structure (e.g. modularity must be retained). The specification of such transformations can be regarded as part of the inital data: what one defines is really an equivalence class, within which a self-organizing system can freely move. (A mathematical study linking structures with structure-preserving transformations is of course called for.)

Such transformations within its equivalence class will actually take place and change the structure of the system during its development (e.g. because forced by external influences, or by the growth of the number of individuals in a population model, or of the volume of trade in an economical model...).

All our attention was focused on finding a general criterion that might allow the determination of the population of each level, once the initial data are given. As such we have proposed, and applied to HMSs, a "principle of invariance of response under allowed structural transformations": "response" or "interaction" is measured through mean values, "allowed transformations" are in our example the p-refinements. That is, the "universe" does not know nor does it care about which class a self-organizing system chooses to settle in, within equivalence.

## 6. DETERMINATION OF THE LAW OF DISTRIBUTION

We must seek to determine the distribution function at the different levels so that the mean value $\langle v \rangle$ remains invariant under any refining transformation. Let us consider the case L = 1 and let the number of elements belonging to level h have the following functional form:

$$n_h^{(p)} = n_0^{(p)} f\,(M^{h/p}) \qquad \text{with } f(1) = 1$$

With p = 2, (1) becomes

$$(4) \quad \langle v \rangle = \frac{1 + Mf(M)}{1 + f(M)} = \frac{1 + \sqrt{M}f(\sqrt{M}) + Mf(M)}{1 + f(\sqrt{M}) + f(M)}$$

from which we obtain

$$(5) \qquad f(M) = M^{-1/2}$$

and hence

$$(6) \qquad n_1^{(2)} = n_0^{(2)} M^{-1/4} \qquad n_2^{(2)} = n_0^{(2)} M^{-1/2}$$

With p generic, (4) becomes

$$(7) \quad \langle v \rangle = \frac{1 + Mf(M)}{1 + f(M)} = \frac{\sum_{h=0}^{p} M^{h/p} f(M^{h/p})}{\sum_{h=0}^{p} f\,(M^{h/p})}$$

Since with p = 2 the functional form of f(M) is reduced to a power of the argument of the function, we shall only consider function f(M) of the type $f(M) = M^x$; (7) becomes

$$(8) \qquad \frac{1 + M^{1+x}}{1 + M^x} = \frac{\sum_{h=0}^{p} M^{(h/p)(x+1)}}{\sum_{h=0}^{p} M^{hx/p}}$$

By associate law, (8) is equivalent to

$$(9) \qquad \frac{1 + M^{1+x}}{1 + M^x} = \frac{\sum_{h=1}^{p-1} M^{(h/p)(x+1)}}{\sum_{h=1}^{p-1} M^{(h/p)x}}$$

It is convenient at this point to introduce the new variable $y = M^{x/p}$; we get

$$(10) \qquad \sum_{h=0}^{p-2} y^{p+h} (M - M^{(h+1)/p}) + \sum_{h=0}^{p-2} y^h (1 - M^{(h+1)/p}) = 0$$

(10) is an algebraic equation in y of 2p-p degree, in which the first p - 1 coefficients are all positive and the last p - 1 are all negative; moreover, the coefficient of power p - 1 is identically equal to zero. (10) therefore generally admits 2p - 2 roots. However we shall show that only one of these is real positive and consequently acceptable for our problem. In fact we can easily verify that (10) admits the solution $y = M^{-1/2p}$ and on the other hand, if we divide (10) by the binomial $y = M^{-1/2p}$, we get a polynomial of degree 2p - 3 with all coefficients positive. This of course, does not admit any real positive root. In fact, we can rewrite (10), arranged in decreasing power order, as follows

$$(11) \qquad \sum_{h=0}^{p-2} \left[ y^{2p-2-h} (M - M^{1-(h+1)/p}) + y^{p-2-h} (1 - M^{1-(h+1)/p}) \right] = 0$$

and dividing the first member of (11) by y - M we get the following equation of degree 2p - 3

$$(12) \qquad \sum_{k=0}^{p-1} \left[ y^{2p-3-k} \sum_{h=0}^{k} (M - M^{1-(k-h+1)/p}) + \right.$$
$$\left. + \; y^{p-2-k} \sum_{h=k}^{p-2} M^{-(k-h+1)/p} (M - M^{(h+1)/p}) \right] = 0$$

whose coefficients are all, at a glance, positive, since the module of the hierarchical system is a number greater than one.

With the solution found we get that the distribution function at the different levels is

$$n_h = n_0 M^{-h/2}$$

It is quite remarkable that the solution found holds good for $p \to \infty$. In this case the number of levels diverges and, if we get the levels obtained by p-refinement, starting from the original situation where the levels are only two, $l = 0$ and $l = 1$, corresponding to the points of a cartesian axis of co-ordinate $h/p$, we can see that for $p \to \infty$, the levels are densely ordered, as are densely arranged the points of real co-ordinates on the interval (0,1). The value function corresponding to the level of z-coordinate is

$$v(z) = M^z$$

and the density distribution function at level z is

$$\rho(z) = \rho_0 M^{-z/2}$$

In fact, if we require that for $p \to \infty$ the mean value remains constant, (9) becomes

$$(13) \quad \sqrt{M} = \lim_{p \to \infty} \frac{\sum_{h=1}^{p-1} M^{(h/p)(x+1)}}{\sum_{h=1}^{p-1} M^{(h/p)x}}$$

$$(14) \quad \sqrt{M} = \lim_{1/p \to 0} \frac{-1 + \dfrac{M^{x+1} - 1}{M^{(x+1)/p} - 1}}{-1 + \dfrac{M^{x} - 1}{M^{x/p} - 1}}$$

The second member of (14) is an indeterminate form which, with the usual methods, gives the following results for $1/p \to 0$

$$(15) \quad \sqrt{M} = \frac{x}{x+1} \; \frac{M^{x+1} - 1}{M^x - 1}$$

and we shall therefore have to find the root of the equation

$$(16) \quad \frac{M^{x+1} - 1}{x+1} = \sqrt{M} \; \frac{M^x - 1}{x}$$

It is easy to verify that (16) admits the root $x = -\frac{1}{2}$ and that moreover it is the only intersection of functions.

We can obtain the same result with another method, i.e. we can show that for a continuous distribution of levels, to which is associated a value function $v(z) = M^z$, with z variable between the lowest level $\ell = 0$ and the highest level $\ell = L$, the distribution density function $\rho(z)$ is proportional to $M^{z/2}$ subject to the condition that the mean value of the set is equal to

$$\langle v \rangle = M^{L/2}$$

In fact we have

$$(17) \quad \langle v\rangle = M^{L/2} = \frac{\int_0^L \rho(z)v(z)dz}{\int_0^L \rho(z)dz} = \frac{\int_0^L \rho(z)M^z dz}{\int_0^L \rho(z)dz}$$

from which

$$(18) \quad \int_0^L \rho(z)M^z dz = \sqrt{M^L}\int_0^L \rho(z)dz$$

After a double derivation with respect to L, we get

$$(19) \quad \rho'(L) + \rho(L)\log\sqrt{M} = 0$$

whose solution is the folowing

$$(20) \quad \rho(L) = \rho(0)\sqrt{M^{-L}}$$

## 7. THE MONETARY SYSTEM

As the simplest model of Hierarchical Modular System, we will consider the monetary system [6], intended as the complex of means used for the exchange of goods and services. HMS's should be regarded as the crudest possible models of self-organizing systems. It was therefore rather surprising to find a realistic situation which is described by them quite satisfactorily. The evidence was first provided by an interesting study by J.C.Hentsch [7], in which a penetrating, empirical analysis was made of monetary circulation in various countries of the world, in the attempt to find some regularities.

All over the world money is organized into a limited number of units that recalls the "levels" of a hierarchical system; the highest level corresponding to the unit with the highest value. The structuring of money into a limited, and usually small, number of distinct units is in itself not without significance; it points to a general principle of "quantification" acting here as a method of assuring the stability of a system with respect to external fluctuations and perturbations. In fact, the use of money of a "continuous" value (for example, gold dust) entails enormous problems (from those of storage and transport to those related to certainty of its value) that do not arise with quantification.

We will discuss three aspects that emerge when "money" is considered as a self-organizing hierarchical modular system: (a) which are the best values to assign to the various coins or notes; (b) is there a criterion that allows the determination of the best distribution of money at each level; (c) how to recognize a monetary system affected by inflation.

The criterion chosen to measure the usefulness of the various methods of organization of a monetary system is the "principle of least effort": the sum necessary for the exchange of goods must be formed on the average by as few coins of various values as possible. It therefore follows, as will be seen in the next section, that the most economical choice is that in which money is structured as a hierarchical modular system.

The answer to the second issue is based on the possibility of deriving a principle of "invariance" for monetary systems. This principle arises quite naturally from the analysis of the interaction between the monetary system and the reality of the economic universe in which the money is used. In fact, the average value in a monetary system depends on the distribution of the coins at the various levels; in conditions of equilibrium, it cannot be arbitrary, but must equal the average price or cost of goods and services; it depends on factors such as cost of labour, availability of natural resources, political organization, and the level of technology of the community,that are outside the structural conditions of the monetary system. The characterizing elements of the monetary system itself are the module, the number of levels and the law of distribution of the money into the various coins and notes.

The latter observation leads us to formulate a principle of invariance under "refining transformation": the law of distribution of the coins must be such that the average value of the monetary system (equal to the average price of the exchanged goods) must remain unchanged regardless of any alteration (= refinement) of the module. In section 6 we deduced the law of distribution that specifies how many coins for each level must stay in circulation to satisfy the above-mentioned criterion. This criterion is evidently derived from conditions of equilibrium in the interaction between the monetary system and the economic universe.

Our results agree well with empirically observed data gathered in about ten different countries. The comparison is particularly significant as the statistical data collected represent a cumulative experience of various civilizations, that has unanimously led to the same empiric law of distribution i.e. the law we derive here from the principle of invariance.

Thus we reduce the subject to a level that may be defined as"static" or "thermostatic" as it refers to situations of equilibrium between the monetary system and the goods and services. However, in reality, these conditions of equilibrium do not remain unchanged; inflation can cause the nominal value of the exchanged objects to increase (and hence there is no longer equality between the average value of the monetary system and the average cost). It must be remembered that the economic, financial and political organization modifies the monetary system and the effective value of the bank-notes in real terms. It is therefore necessary to understand the processes of reorganization of monetary systems induced by inflation. Here again, the average price, being equal to the effective average value of the monetary system, takes into account the reorganization process undergone by money, i.e. the disappearance of coins of a low value and the simultaneous emission of coins of higher value.

A final note concerns a rule that will be followed hereafter: as we are interested only in integer numbers, any real number is to be rounded to the nearest integer at the corresponding level, so as not to enter into irrelevant problems of discrete mathematics, that would only complicate matters.

## 8. THE MONETARY SYSTEM AS A HIERARCHICAL MODULAR SYSTEM

We should first evaluate why money is minted in various values and what are the best values to assign to the various coins. First of all, let us establish, in order to evaluate the greatest or the smallest utility of the possible organizations of monetary systems, a criterion of economy based on the need to be able to buy goods with a low number of coins. The argument may further be clarified by evaluating in various cases the average number $\bar{n}$ of coins necessary to form any number between the minimum value, for example one "lira", and a maximum value corresponding to S "lire".

It is obvious that this average number cannot be below one. The value $\bar{n} = 1$ means that all denominations are issued, that is there are as many levels as there are numbers between 1 and S. Such a monetary system is too complicated as it requires a coin for every possible number and, moreover, it would be too awkward to use. It is equally clear that the maximum value of $\bar{n}$ corresponds to the choice of issuing one single coin, that of 1 lira. In this case

$$\bar{n} = (S + 1)/2$$

and the number of levels is L = 1. It is evident, however, that also in this case the monetary system is too complicated to use. We are faced with two conflicting needs: the minimum number of the levels should correspond to the situation in which $\bar{n}$ is maximum and, viceversa, the lowest value of $\bar{n}$ should correspond to the situation in which L is maximum.

The problem is better posed if reformulated as follows: if L is the number of levels, i.e. the number of different values to give the coins, which values must be chosen for these L coins in order to minimize $\bar{n}$?

It is clear that the average number of coins necessary to form all numbers between 1 and S depends on what values are given to each coin. We will show that the average number of coins necessary to form all the numbers between 1 and S becomes minimum if the monetary system is modular, that is if the value associated to the h level is determined by a law of the type $v_h = cM^h$ (M is the module, c a constant).

From this point of view, the Italian monetary system e.g.,that is all the coins and bank-notes circulating in Italy, is well organized. The number of levels is L = 14; in fact, it is possible to regroup the money into 14 subgroups: first, all the 5 lire coins; second, those of 10 lire; third, those of 20 lire; fourth, 50 lire; fifth, 100 lire; sixth, 200 lire and so on until the fourteenth, where we have 100,000 lire bank-notes. The value function of one element is simply the face value of money and the module is $M = \sqrt[3]{10}$. In fact, with this module we have the sequence of values:

1; 2.15; 4.64; 10; 21.5; 46.4; 100 ...

that when rounded off to the nearest integer of the corresponding level, well corresponds to the sequence:

1; 2; 5; 10; 50; 100 ...

in actual use in the Italian monetary system (the first two values have become obsolete).

To demonstrate the foregoing points, let us first observe that a generic integer number x falling between 1 and S may be expressed by a combination such as

$$x = a_0 v_0 + a_1 v_1 + a_2 v_2 + \ldots + a_L v_L$$

where $v_0 = 1$ and $v_1, \ldots, v_L$ represent the values of the money at the L + 1 levels and the $a_0, a_1, \ldots, a_L$ coefficients are necessarily integers greater than or equal to zero.

If we arrange the values $v_0 = 1, v_1, \ldots, v_L$ in progressive order and choose them in such a way that the ratio

$$\frac{v_{h+1}}{v_h} = n_h$$

with $n_h$ an arbitrary integer, to get the numbers from $v_0 = 1$ to $v_1 = 1$, we need in total $\frac{1}{2}v_1(v_1 - 1)$ coins of value $v_0$. If the numbers from $v_1$ to $v_2 - 1$ are to be made up in the most economical manner (that is, where possible, not using $v_1$ coins worth $v_0$ in place of one coin worth $v_1$), to form the numbers between 1 and $v_2 - 1$, a total of $\frac{1}{2}v_2(v_2/v_1 - 1)$ coins are necessary.

That is, to express all the integers up to $v_L - 1$ we need:

$$\frac{1}{2}(v_1 - 1)v_1 \frac{v_2}{v_1}\frac{v_3}{v_2} \ldots \frac{v_L}{v_{L-1}} = \frac{v_L}{2}(v_1 - 1) \qquad \text{coins worth } v_0$$

$$\frac{1}{2} v_1 \frac{v_2}{v_1} \ldots \left(\frac{v_{r+1}}{v_r} - 1\right) \ldots \frac{v_L}{v_{L-1}} = \frac{v_L}{2}\left(\frac{v_{r+1}}{v_r} - 1\right) \text{coins worth } v_r$$

$$\frac{1}{2} v_1 \frac{v_2}{v_1} \ldots \frac{v_L}{v_{L-1}}\left(\frac{v_L}{v_{L-1}} - 1\right) = \frac{v_L}{2}\left(\frac{v_L}{v_{L-1}} - 1\right) \qquad \text{coins worth } v$$

and in total we need

$$\frac{v_L}{2} \cdot v_1 + \frac{v_1}{v_2} + \ldots + \frac{v_L}{v_{L-1}} - L$$

coins. The minimum number of coins necessary is therefore determined by the single ratios $v_{r+1}/v_r$ and is achieved ($v_L$ and L are fixed by hypothesis) when the sum

$$v_1 + \frac{v_2}{v_1} + \dots + \frac{v_L}{v_{L-1}}$$

is minimum. With fixed product of these ratios (being equal to $v_L$), the search for L numbers whose sum is minimal in correspondence to an assigned product. Such a situation arises when all the terms of the sum are equal, that is

$$v_{r+1}/v_r = M = (v_L)^{1/L}$$

Having fixed the number $v_L = M^L$ and the L number of levels, the minimum number of coins which are necessary to obtain all the integers between zero and M - 1 is

$$N = \frac{L}{2}(M - 1)v_L$$

This type of system is called "modular with module M". We may therefore conclude that a modular system minimizes the number of tokens necessary to obtain all the numbers between zero and $v_L - 1$, with respect to any other division (in L + 1 levels).

Regarding the choice of the most convenient module, once again we can calculate the average number of coins needed to form the numbers in an assigned congruous interval of values, for instance, between 1 and $M^L - 1$. The tokens at our disposal are $M^0 = 1, M, M^2, \dots, M^L$. We will calculate $\bar{n}$ in the hypothesis that each coin may not be used more than necessary (therefore a maximum of M - 1 times). Thus, if we take as an example the Italian monetary system, to make up 1350 lire, we have a 50 lire coin, a 100 lire coin, a 200 lire coin and a 1000 lire note; 7 coins of 50 lire each and a 1000 lire note is not considered "good". The average number of coins necessary to form the numbers in the interval of values 1, $M^L - 1$ is given by

$$\bar{n} = \frac{L(M - 1)}{2}$$

(M - 1)/2 being the average number of coins for each value necessary to make up the numbers in the considered interval. It follows that $\bar{n}$ is a growing function of the module. This implies that the most convenient monetary system module is M = 2. However, besides the minimum condition on $\bar{n}$, at least two other elements should be considered.

a) the first is that for the countries using the decimal system, the module must be a power of ten to make mental calculations easy;

b) the other is that the module cannot be too small; otherwise to cover an assigned interval of 1, $v_{max}$ values, it is necessary to increase the number of levels and therefore of individual coins, with all the practical inconveniences that this causes.

The value $M = \sqrt[3]{10} = 2.25$ (not very distant from the integer 2) found in the Italian monetary system well fits all the necessary conditions.

## 9. INVARIANCE PRINCIPLE AND DISTRIBUTION FUNCTION OF THE COINS

Let us now consider a hierarchical modular system composed of L + 1 levels, and of module M. The module may be changed to introduce other levels into the system besides the already existing L + 1. In particular the number of the levels passes from L + 1 to pL + 1 with the transformation $M \longrightarrow M' = M^{1/p}$ with p integer. Let us say that the system has undergone a refinement of the order p. For instance, let us consider the hierarchical system formed by coins of one lira that constitute the zero level and by 10 lire coins that make up level one; in this system the module is M = 10. If a 3-refinement is carried out the module becomes $M' = \sqrt[3]{10}$ and the levels are now

level 0 to which belong coins of 1 lira
level 1 to which belong coins of 2 lire
level 2 to which belong coins of 5 lire
level 3 to which belong coins of 10 lire

The total value and the average value of a monetary system obviously depend on how the N coins are distributed at the L + 1 levels. The <v> average value of an element of a monetary system cannot be an arbitrarily chosen quantity; it is fixed by the average value of all the goods exchanged in an assigned time interval by the community which uses that particular monetary system; it depends on the particular exchange needs of the community, not on the structure of the monetary system, and it cannot depend on the choice of the module of the hierarchical system.

We shall therefore formulate the principle of invariance <v> for refining transformations: the mean value of an element of a monetary system is invariant under under $M \longrightarrow M^{1/p}$ refinement, regardless of what p is. This is a very important aspect, as the discovery of a principle of invariance has always been a powerful analytic tool in all fields of scientific research.

From this principle of invariance, using the results of sections 5 and 6, we derive the distribution function of the coins at the levels of the hierarchical system, compatible with the expressed principle of invariance.

$$n_h = n_0 M^{-h/2} = \frac{n_{h-1}}{\sqrt{M}} \qquad (21)$$

The number of elements belonging to the level h equals the number of elements belonging to the level immediately below divided by the square root of the module.

## 10. COMPARISON WITH SOME ACTUAL MONETARY SYSTEMS

In section 9 we have outlined the law of distribution of different coins in the various levels of the monetary system. If the principle of invariance by refinement is correct, there must be a strict correspondence with the distribution of coins in the systems in actual use.

To make this comparison we observe that, with the law of distribution (21), the total value of the money in circulation belonging to an assigned level h, that we will indicate with $A_h$, is given by

$$A_h = n_h v_h = n_0 M^{h/2} = n_0 \sqrt{v_h}$$

from which

$$\log A_h = \log n_0 + \tfrac{1}{2} \log v_h$$

If we transfer onto logarithmic paper $x = \log v_h$ on the abscissa and $y = \log A_h$ on the ordinate, the points should be distributed on a 1/2 angular coefficient line.

Figure 2 shows the data relevant to the money circulating in Switzerland at the end of 1969 and 1971. It is obvious that when the low value coins were minted in precious metal (1969) the money in circulation was above that expected by the law of distribution; this was due to hoarding by private citizens. The data for 1971 refer to a situation in which the coins were no longer minted in silver, and as can be seen from the figure, the circulation has become almost normal.

Figure 3 shows the data relative to money in circulation in Italy in June 1978. The differences with respect to the law of distribution of notes are bigger than those found in other countries. In particular, an increase of 2,000, 5,000 and 20,000 lire notes is necessary. This would automatically bring about a regulation of the 10,000 lire notes in circulation.

The other figures show the data corresponding to the following countries: Finland, U.S.A., Yugoslavia, West Germany, Austria, England and Sweden. All cases show substantial confirmation of the law of monetary distribution deduced here.

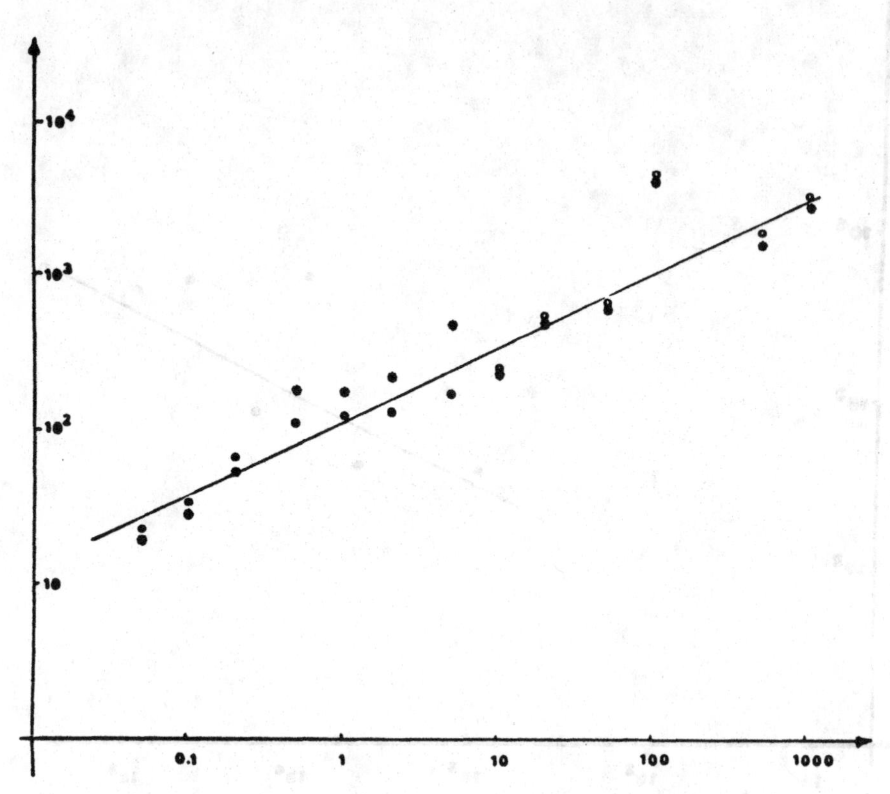


Fig. 2 * Svizzera 31.12.69
o Svizzera 31.12.71

Ordinates are in million Franks.
Two important effects may be seen: the absence of 200 franc coins is the cause of a marked increase in circulation of 100 franc coins. Moreover, as soon as an external perturbation ceases (in this case the hoarding of silver coins) the monetary circulation tends towards the canonical distribution (compare the data for coins of 0.5, 1,2, and 5 francs).

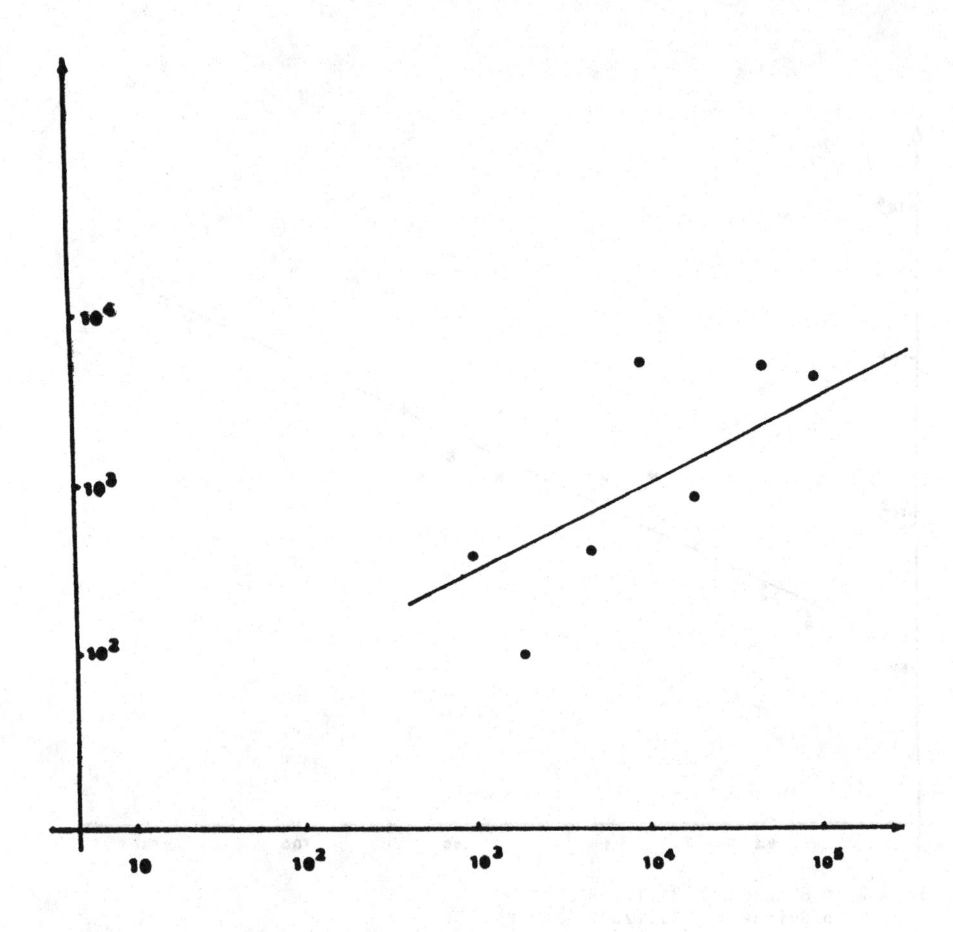


Fig. 3 Italy - Italia 30-6-78
Ordinates are in billion lire.

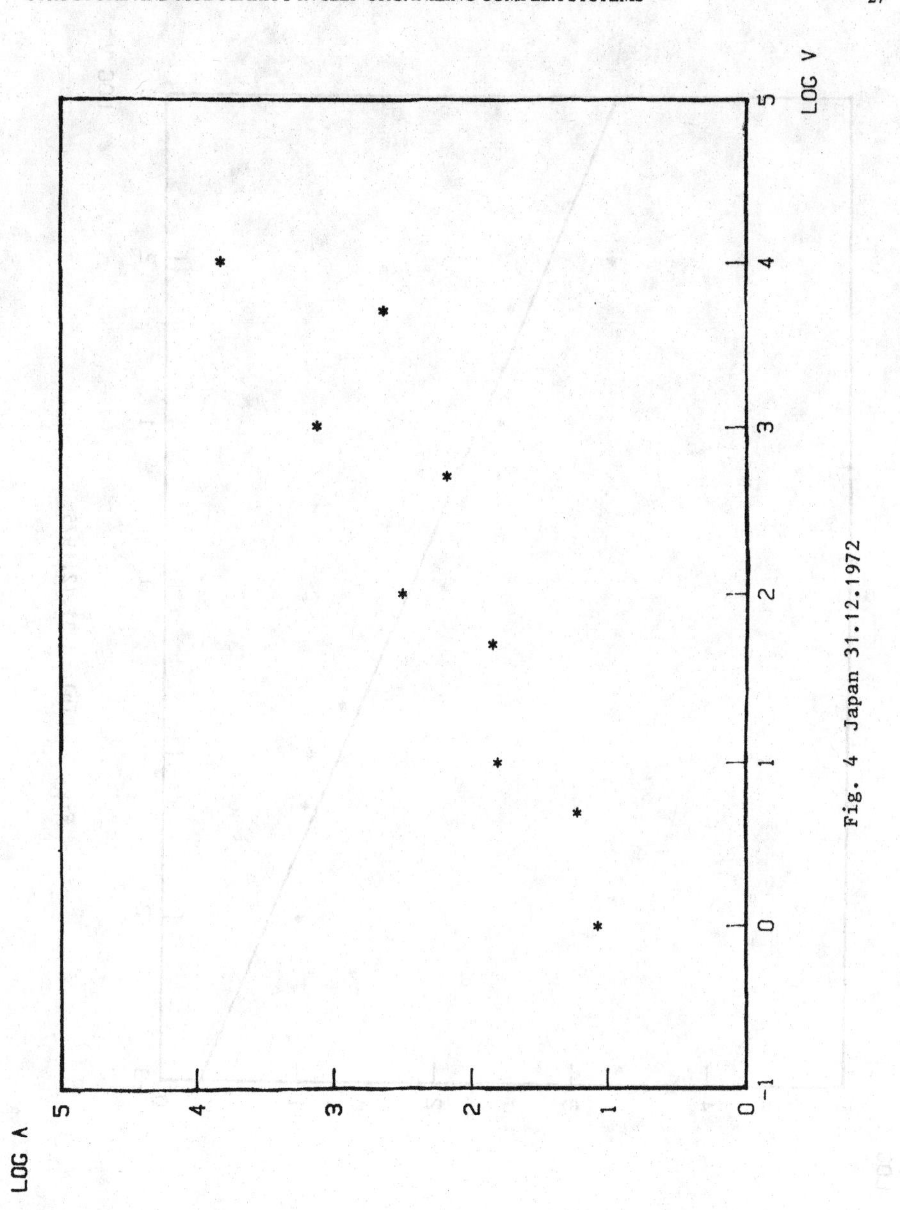


Fig. 4 Japan 31.12.1972

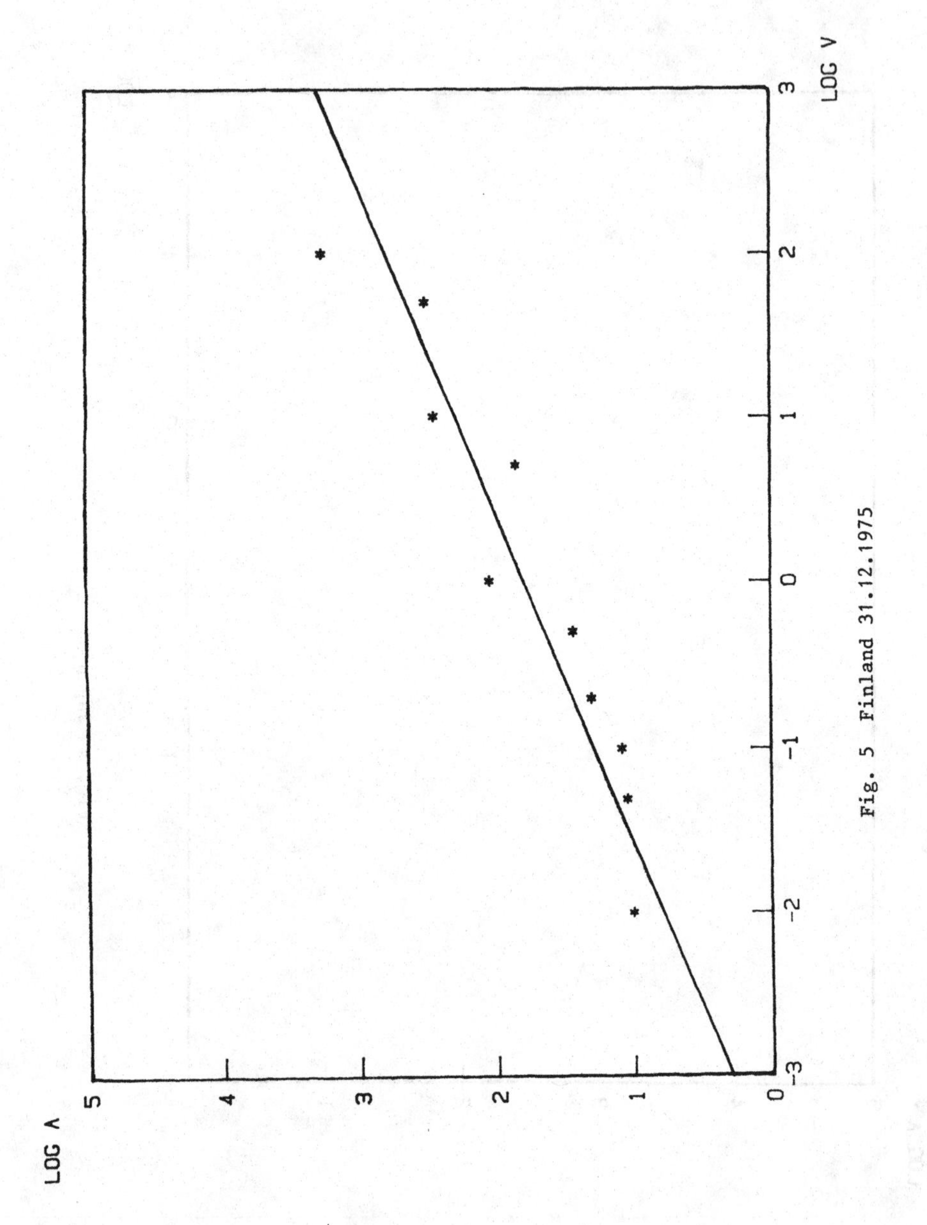


Fig. 5 Finland 31.12.1975

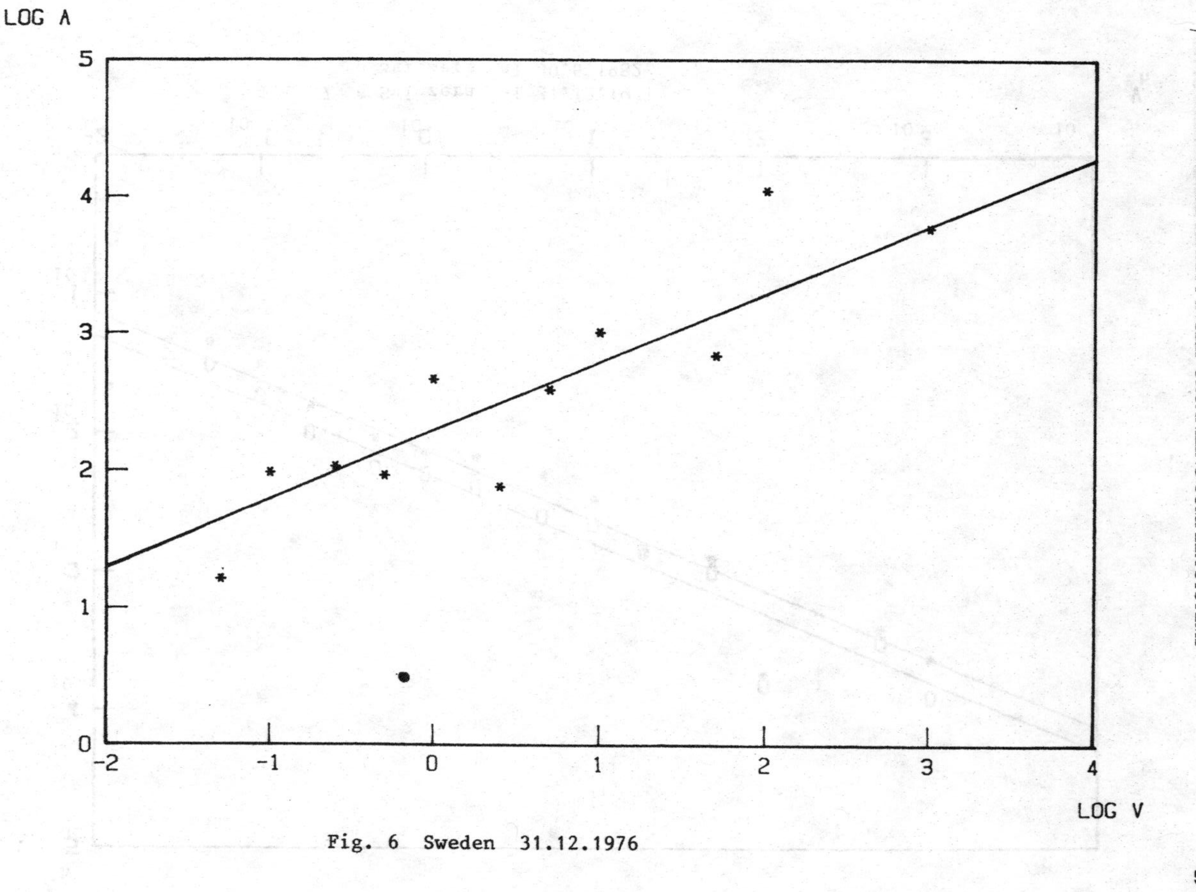


Fig. 6 Sweden 31.12.1976

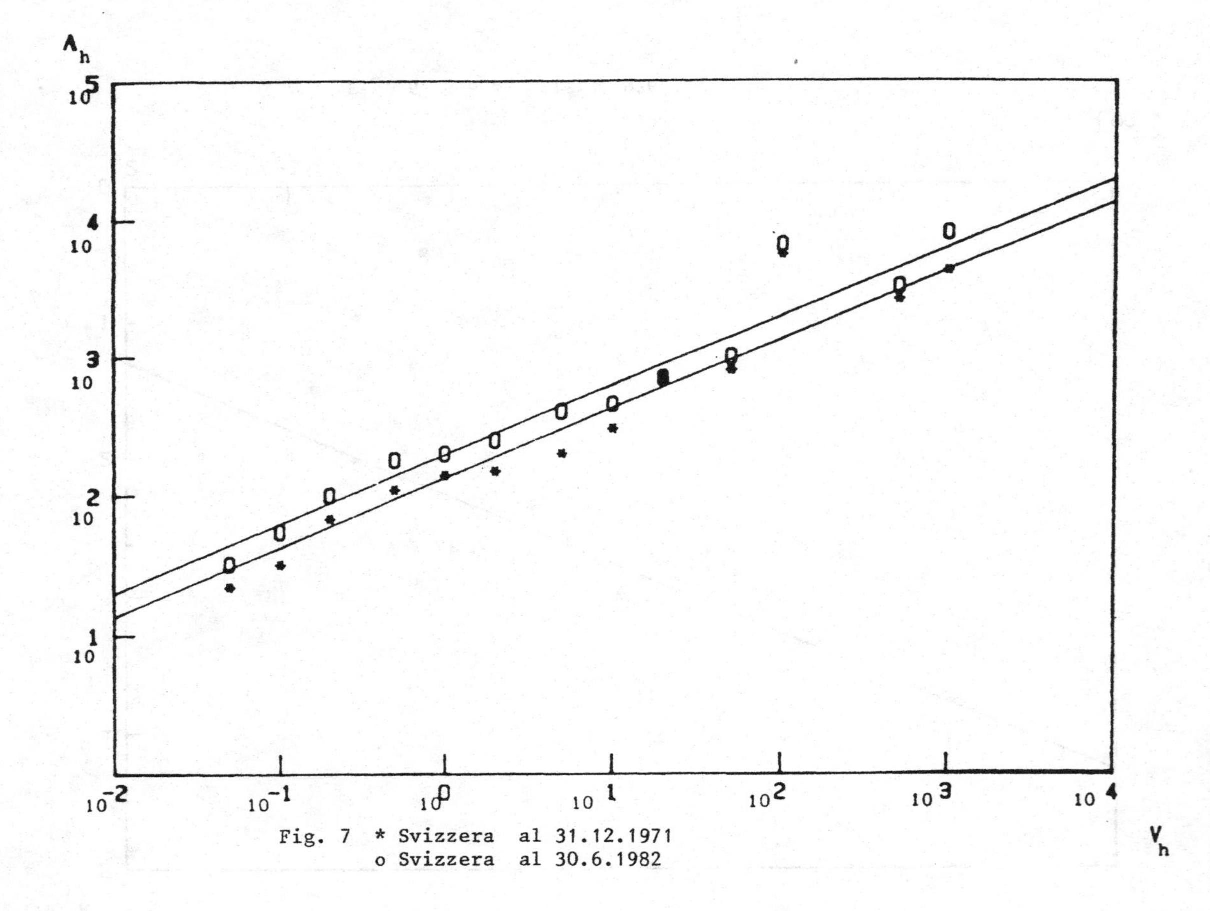


Fig. 7 * Svizzera al 31.12.1971
o Svizzera al 30.6.1982

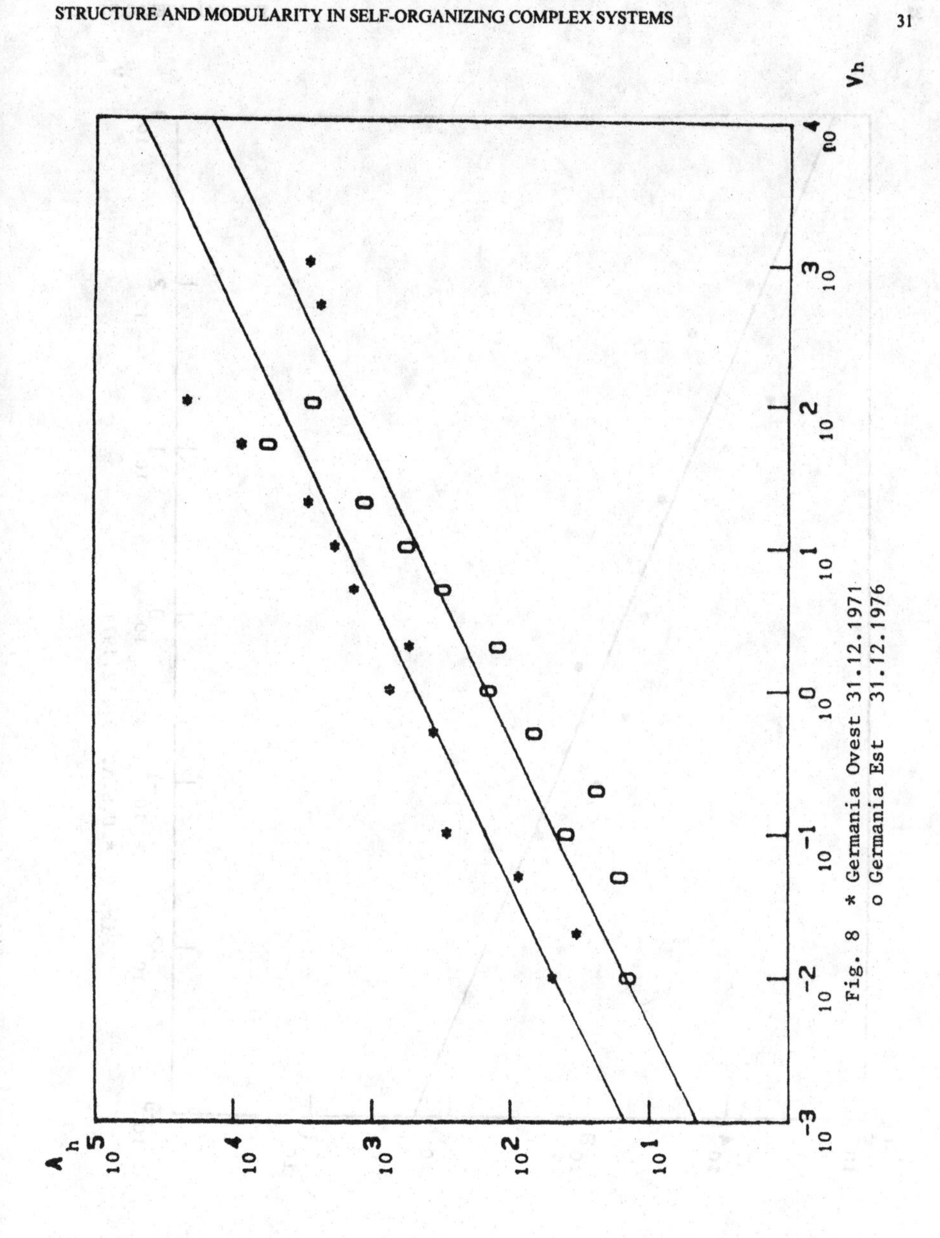


Fig. 8 * Germania Ovest 31.12.1971
o Germania Est 31.12.1976

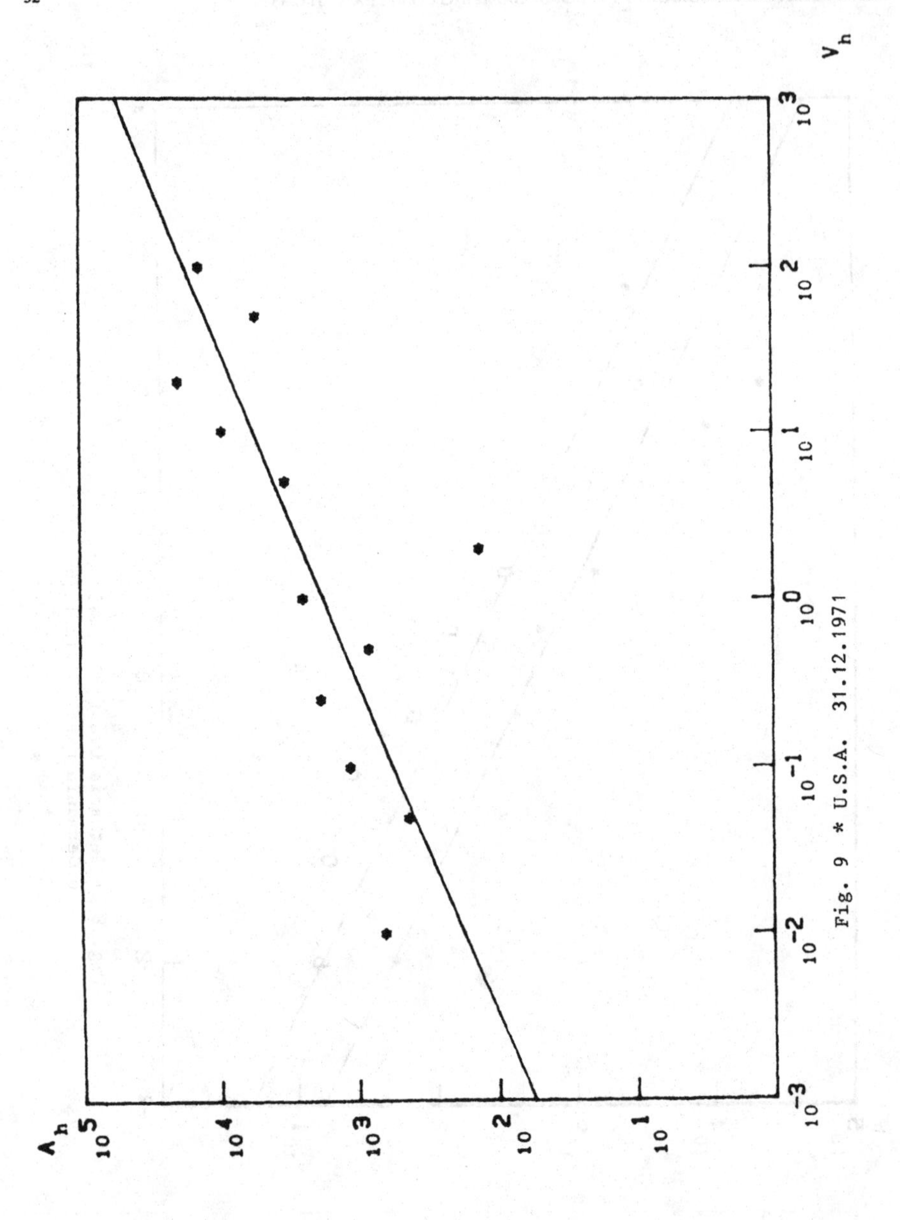


Fig. 9 * U.S.A. 31.12.1971

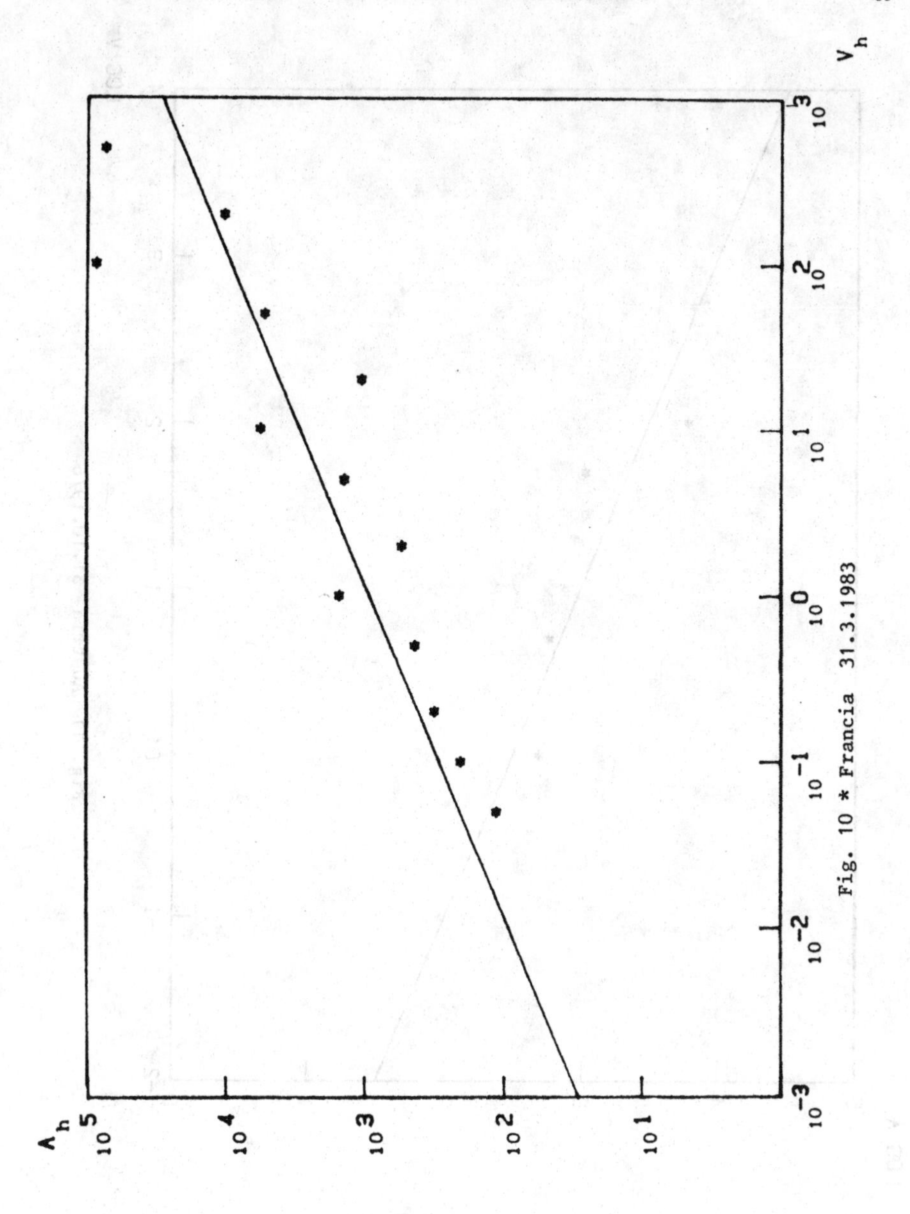


Fig. 10 * Francia 31.3.1983

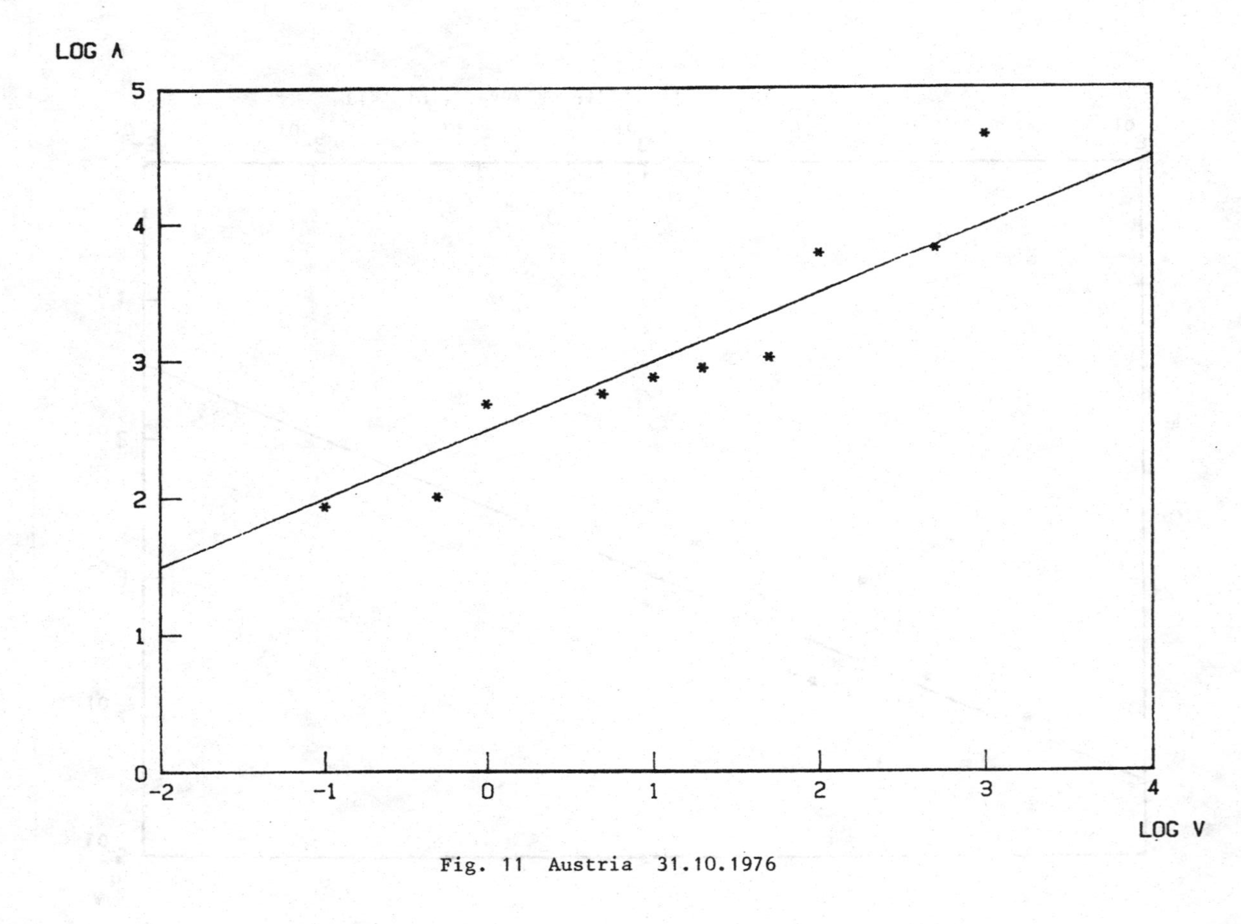


Fig. 11 Austria 31.10.1976

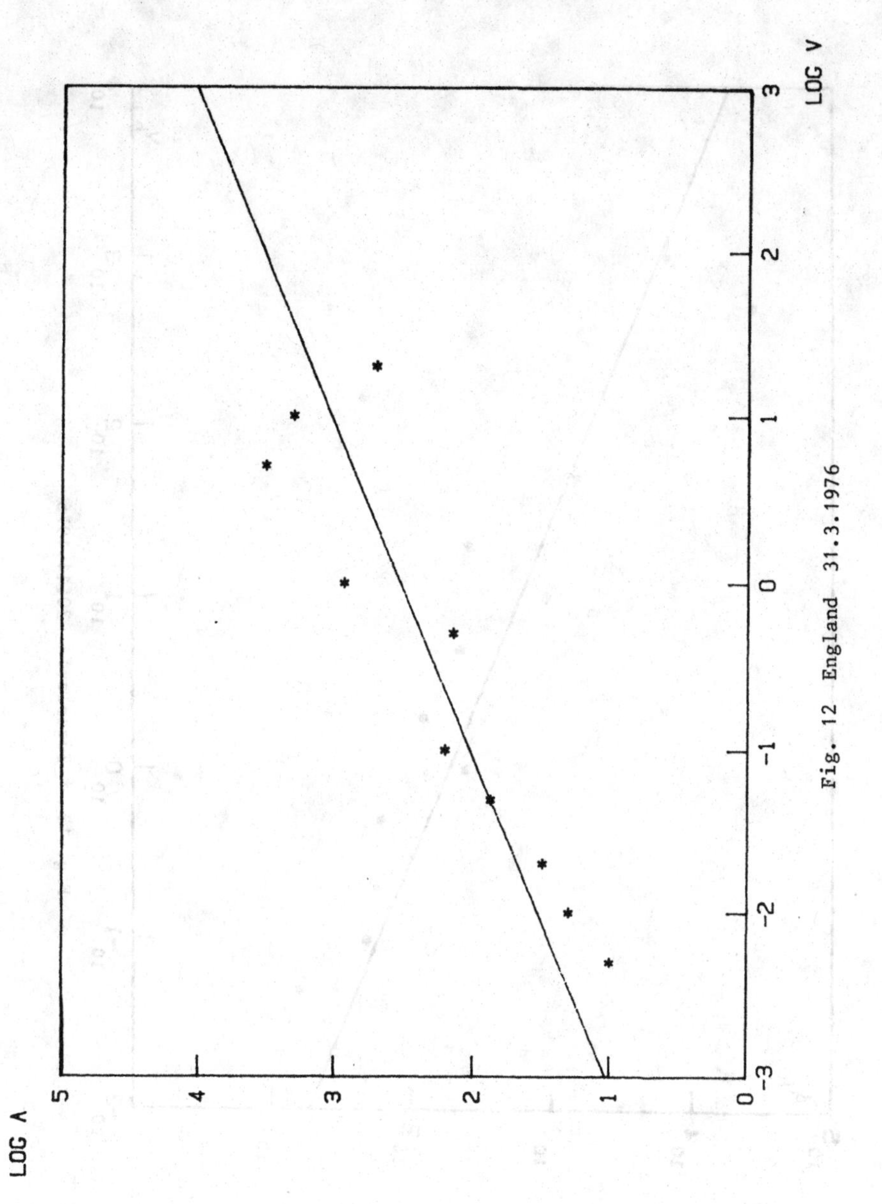


Fig. 12 England 31.3.1976

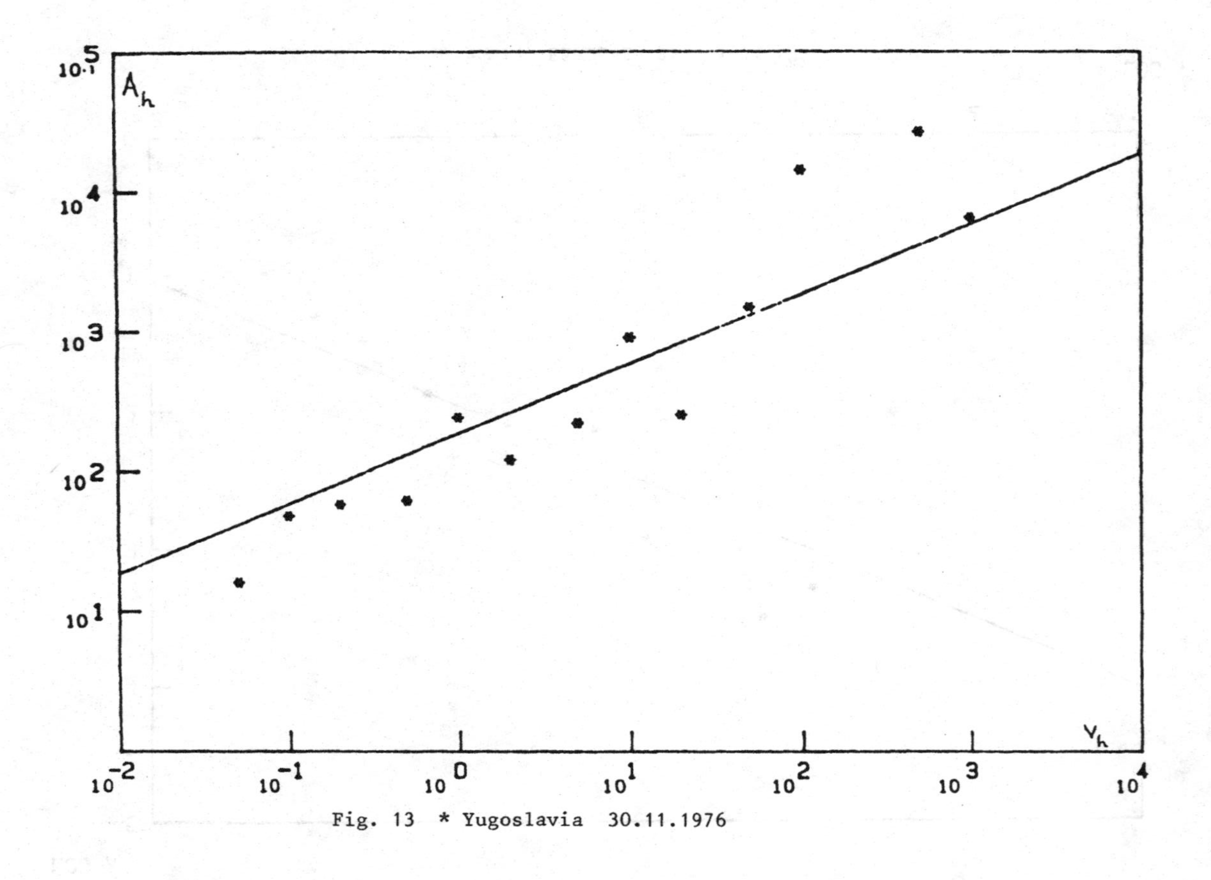


Fig. 13 * Yugoslavia 30.11.1976

## 11. REORGANIZATION OF THE MONETARY SYSTEM DUE TO INFLATION

It should be understood that we are not concerned with the causes of inflation, but only on how the processes of inflation influence the evolution of monetary systems. In the previous sections we have seen that the law of distribution of money in the various levels derives from the principle of invariance by refining transformations, provided the average value of goods $\langle v \rangle$ equals the average value of the monetary system, that is equal to $\sqrt{M^L}$. Inflation causes the intrinsic value of the goods exchanged to increase with time, therefore the average value $\langle v \rangle$ cannot remain constant but must become a monotonic, not decreasing function of time: $\langle v \rangle = v(t)$. If, therefore, we start from a situation in which there is equilibrium between the monetary system and the economic system, with time this equilibrium is upset by the process of inflation and we have

$$v(\tau + t) > v(\tau) = M^L$$

To re-establish the equilibrium, the monetary system reorganizes itself following the principle of "least effort". There are, in fact, three possible responses from the monetary system:
a) the process of re-equilibration of the monetary system with the external world comes about by a suitable change of the module, that at $\tau + t$ moves from value M to value $M' > M$ so as to have $v(\tau + t) = \sqrt{M'^L}$;
b) the module does not change, but the number of levels increases from L to L' by the emission of notes of a higher value so that we have $v(\tau + t) = \sqrt{M'^{L'}}$;
c) neither the module nor the number of levels changes, but there is a process of re-equilibration by the emission of new notes with a higher value and at the same time by the disappearance of a corresponding number of notes of a lower value, so that we have $v(\tau + t) = c(\tau + t)\sqrt{M^L}$ with $c(\tau) = 1$ and $c(\tau + t) = \sqrt{M^z}$, z being equal to the number of notes fallen into disuse in consequence of the process of reorganization.

It is obvious that the third type of response is the most "economical" and therefore the one generally adopted in the reality of monetary systems. In fact, the change of the module implies the substitution of all coins and notes in circulation with those corresponding to the new series of values determined by module M'; in addition it would cause a further difficulty determined by the change in the base of numbering. All this requires a cost and an effort so great that this type of reorganization process of monetary systems affected by inflation is not feasible. On the other hand, the increase in the number of levels contrasts with the requirements of simplicity and ease of use that comes about only with a sufficiently low number of coins. It should be noted that in fact the various existing monetary systems are composed of a number of levels ranging from 10 to 14, usually divided in turn into two equally numerous subgroups each containing from 5 to 7 levels, one corresponding to the number of metal coins, the other to the number of paper notes. The latter alternative leaves the structural elements of the monetary system, i.e. the module and the number of levels unchanged, and this process is not subjected to the disadvantages and difficulties of the first two cases; the principle of "least effort" gives the model of reorganization followed in reality by monetary systems.

## 12. DISTRIBUTION OF A POPULATION ON A TERRITORY

We examined, as a first example of an HMS, the monetary system, for which the number of levels and the values of elements in each level are clearly known a priori. A different situation occurs for the system we are going to analyse as a second example, i.e. that whose elements are constituted by the urban settlements existing over a given territory. For this system the hierarchical structure is no more clearly evident a priori, because both the number of levels in which the system is structured and the value of an element in each level are not known. Therefore the problem of determining the distribution of the elements between the different levels can be faced in this case (as in many other natural systems devoid of an explicit given hierarchical structure), only after finding the criteria allowing an objective assignment of the number of levels and of the value associated to each of them.

For instance, in fig.4 we report the population distribution in Campania in municipalities of up to 20,000 inhabitants; the only relevant datum is the rapid decrease in the number of municipalities belonging to a given interval with respect to the increase of inhabitants; this corresponds to the well known Pareto law of exponential decrease.

On the other hand, it is intuitively clear that the population is distributed according to a structure with levels, in the sense that in higher levels one puts the cities with a higher number of inhabitants, in which a more complex and efficient organization allows each inhabitant to have on average a better quality and a higher standard of life.

The German geographer W.Christaller was the first, to our knowledge, who recognized, in his treatise "Die zentralen Orte in Süddeutschland" [8] , a hierarchical structure with levels in the distribution of urban settlements. Christaller is the founder of a new approach to human geography that proposes to explain the spatial relations between the municipalities on the grounds of production relations. He founds his research on the proposition that "every economical relation and every economical process have, without exception, a spatial dimension".

According to Christaller, the most relevant peculiarity of the municipalities for the purpose of the survey of their spatial distribution is what he calls "centrality", or vocation to act as attraction pole with respect to a given territorial area. The centrality of a place is a relative notion, because the place can be "central" respect to a given territory, but, at the same time, inside the orbit of a higher central place with respect to a wider territorial frame.

The centrality of a place is in close relation to the goods and services that are produced inside; he distinguishes between "dispersed" or "indifferent" goods and services, that can be supplied either in central places or in non-central ones, and "central" goods and services, supplied only in the central municipality of a given territory.

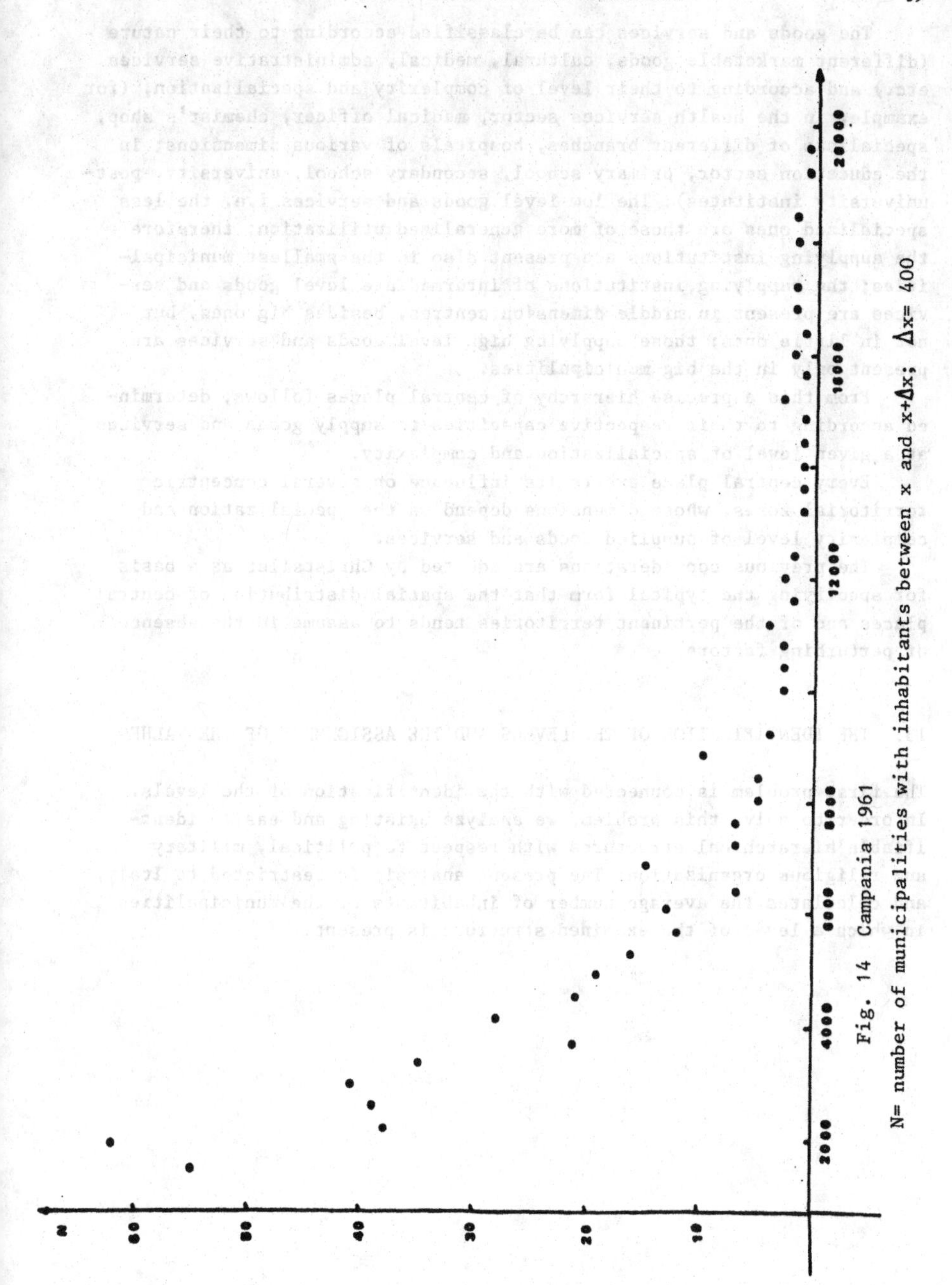


Fig. 14 Campania 1961

N= number of municipalities with inhabitants between x and x+$\Delta$x; $\Delta$x= 400

The goods and services can be classified according to their nature (different marketable goods, cultural, medical, administrative services, etc.) and according to their level of complexity and specialization, (for example, in the health services sector, medical officer, chemist's shop, specialists of different branches, hospitals of various dimensions; in the education sector, primary school, secondary school, university, post-university institutes). The low level goods and services i.e. the less specialized ones are those of more generalized utilization; therefore the supplying institutions are present also in the smallest municipalities; the supplying institutions of intermediate level goods and services are present in middle dimension centres, besides big ones, but not in little ones; those supplying high level goods and services are present only in the big municipalities.

From this a precise hierarchy of central places follows, determined according to their respective capacities to supply goods and services at a given level of specialization and complexity.

Every central place exerts its influence on several concentric territorial zones, whose dimensions depend on the specialization and complexity level of supplied goods and services.

The previous considerations are adopted by Christaller as a basis for specifying the typical form that the spatial distribution of central places and of the pertinent territories tends to assume in the absence of perturbing factors.

## 13. THE IDENTIFICATION OF THE LEVELS AND THE ASSIGNMENT OF THE VALUES

The first problem is connected with the identification of the levels. In order to solve this problem, we analyze existing and easily identifiable hierarchical structures with respect to political, military and religious organization. The present analysis is restricted to Italy, and calculates the average number of inhabitants of the municipalities in which a level of the examined structure is present.

TABLE 1
Average number of inhabitants of the Italian municipalities according to the location of CARABINIERI

| LEVELS | AVER. POPUL. | LOG. AVER. POPUL. |
|---|---|---|
| 0 Municipalities without Carabinieri | 2000 | 3.30 |
| 1 Municipalities with Carabinieri | 5400 | 3.73 |
| 2 Municipalities with Comando Tenenza | 16500 | 4.21 |
| 3 Municipalities with Comando di Compagnia | 35000 | 4.54 |
| 4 Municipalities with Comando di Gruppo | 100000 | 5.00 |
| 5 Municipalities with Comando di Legione | 200000 | 5.30 |
| 6 Municipalities with Comando di Brigata | 570000 | 5.75 |
| 7 Municipalities with Comando di Divisione | 1500000 | 6.18 |
| 8 Municipalities with Comando Generale | 2880000 | 6.48 |

TABLE 2
Average population of the Italian municipalities according to ADMINISTRATIVE HIERARCHY

| LEVELS | AVER. POPUL. | LOG. AVER. POPUL. |
|---|---|---|
| 0 Municipalities not chief town | 4700 | 3.67 |
| 1 Municipalities chief town of Provincia | 101000 | 5.00 |
| 2 Municipalities chief town of Regione | 440000 | 5.64 |
| 3 Capital City | 2880000 | 6.46 |

TABLE 3
Average population of the Italian municipalities according to the location of GUARDIA DI FINANZA

| LEVELS | AVER. POPUL. | LOG. AVER. POPUL. |
|---|---|---|
| 0 Municipalities without G.F. posts | 3600 | 3.58 |
| 1 Municipalities with G.F. posts | 15000 | 4.18 |
| 2 Municipalities with Comando di Tenenze G.F. | 20000 | 4.30 |
| 3 Municipalities with Comando di Compagnia G.F. | 35700 | 4.55 |
| 4 Municipalities with Comando di Gruppo | 91000 | 4.91 |
| 5 Municipalities with Comando di Legione | 157500 | 5.22 |
| 6 Municipalities with Comando di Zona | 575000 | 5.76 |
| 7 Municipalities with Comando di Ispettorato | 1500000 | 6.18 |
| 8 Municipalities with Comando Generale | 2880000 | 6.46 |

TABLE 4
Average population of the Italian municipalities according to the location of IMPOSTE DIRETTE

| LEVELS | AVER. POPUL. | LOG. AVER. POPUL. |
|---|---|---|
| 0 Municipalities without I.D. posts | 4200 | 3.62 |
| 1 Municipalities with uffici distrettuali | 26000 | 4.41 |
| 2 Municipalities with Intendenze di finanze | 100000 | 5.00 |
| 3 Municipalities with Ispettorato Compartimentale | 570000 | 5.70 |
| 4 Municipalities with Direzione generale | 2880000 | 6.46 |

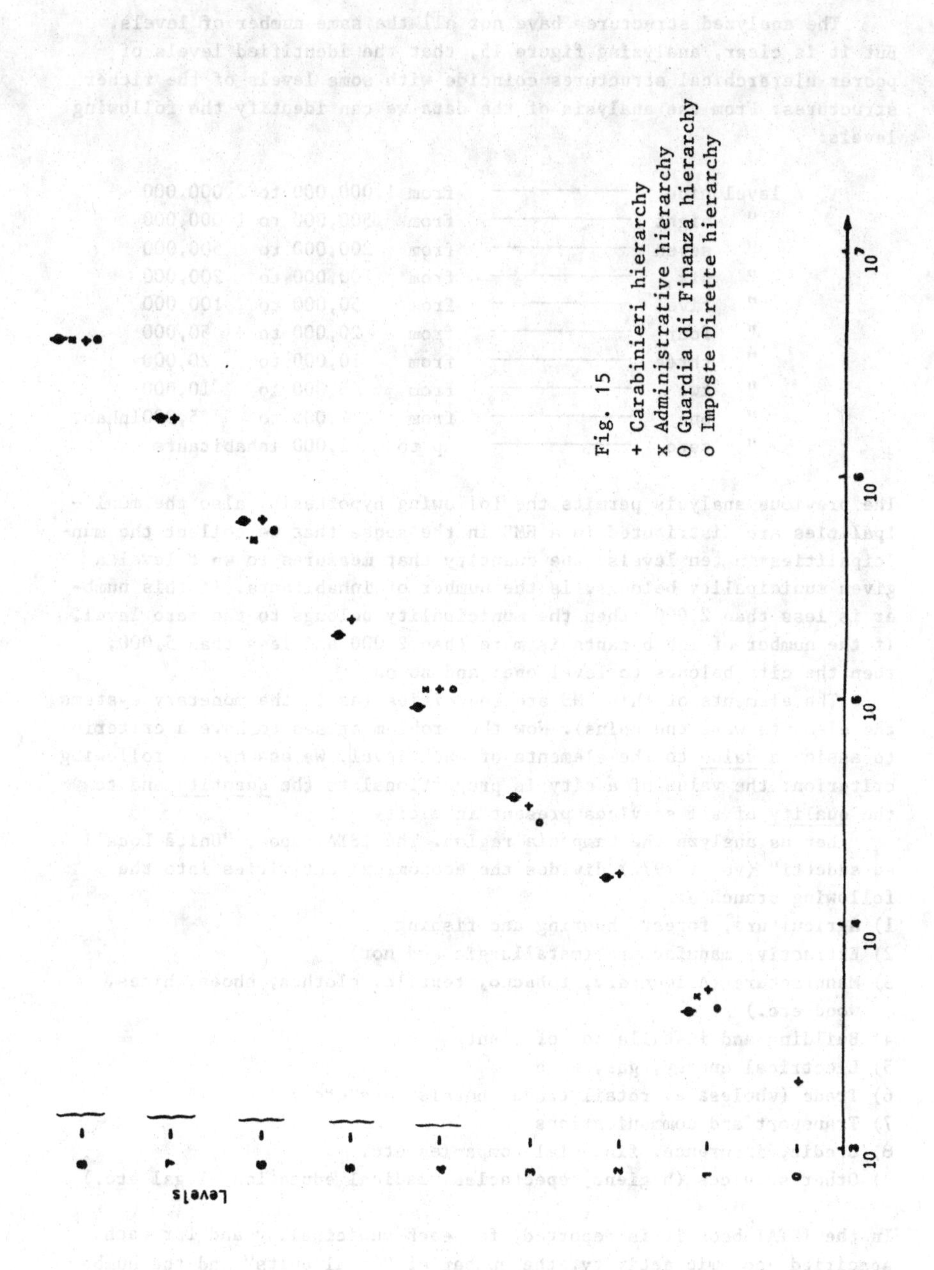
+ Carabinieri hierarchy
x Administrative hierarchy
O Guardia di Finanza hierarchy
o Imposte Dirette hierarchy
Levels
0
1
2
3
4
5
6
7
8
9
$10^3$
$10^4$
$10^5$
$10^6$
$10^7$

Fig. 15

The analyzed structures have not all the same number of levels. But it is clear, analyzing figure 15, that the identified levels of poorer hierarchical structures coincide with some levels of the richer structures. From the analysis of the data we can identify the following levels:

| | | | |
|---|---|---|---|
| level nine | ———— | from 1,000,000 to 2,000,000 |
| " eight | ———— | from 500,000 to 1,000,000 |
| " seven | ———— | from 200,000 to 500,000 |
| " six | ———— | from 100,000 to 200,000 |
| " five | ———— | from 50,000 to 100,000 |
| " four | ———— | from 20,000 to 50,000 |
| " three | ———— | from 10,000 to 20,000 |
| " two | ———— | from 5,000 to 10,000 |
| " one | ———— | from 2,000 to 5,000inhab. |
| " zero | ———— | up to 2,000 inhabitants |

The previous analysis permits the following hypothesis: also the municipalities are distributed in a HMS in the sense that we collect the municipalities in ten levels. The quantity that measures to what level a given municipality belongs, is the number of inhabitants. If this number is less than 2,000, then the municipality belongs to the zero level. If the number of inhabitants is more than 2,000 and less than 5,000, then the city belongs to level one; and so on.

The elements of this HMS are the cities (as in the monetary systems the elements were the coins). Now the problem arises to have a criterion to assign a value to the elements of each level. We assume the following criterion: the value of a city is proportional to the quantity and to the quality of all services present in a city.

Let us analyze the Campania region. The ISTAT book, "Unità Locali ed addetti" (vol.1,1972) divides the economical activities into the following branches:

1) Agriculture, forest, hunting and fishing
2) Extractive manufacture (metallurgic and not)
3) Manufacture (Alimentary, tobacco, textile, clothes, shoes, hides, wood etc.)
4) Building and installation of plants
5) Electrical energy, gas, water
6) Trade (wholesale, retail trade, hotels bars etc.)
7) Transport and communications
8) Credit, insurance, financial companies etc.
9) Other services (hygiene, spectacles, medical education, legal etc.)

In the ISTAT book it is reported, for each municipality and for each specified economic activity, the number of "local units" and the number of workers.

Now we calculate, for each level, the average number of "local units" of the municipality belonging to the given level and moreover the average number of workers. We obtain

**Average number of "local units" of the municipalites belonging to the given level**

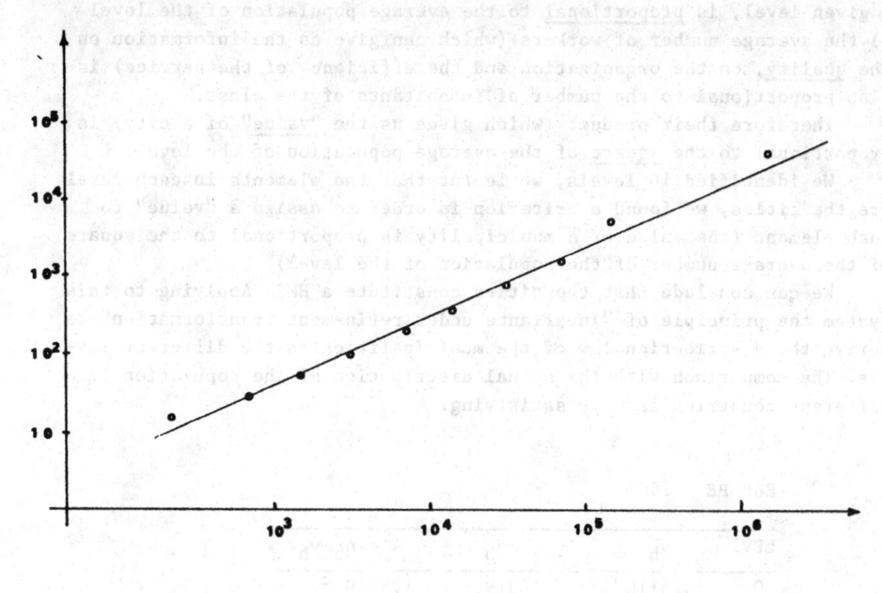


Fig. 16

**Average number of workers of the municipalities belonging to the given level**

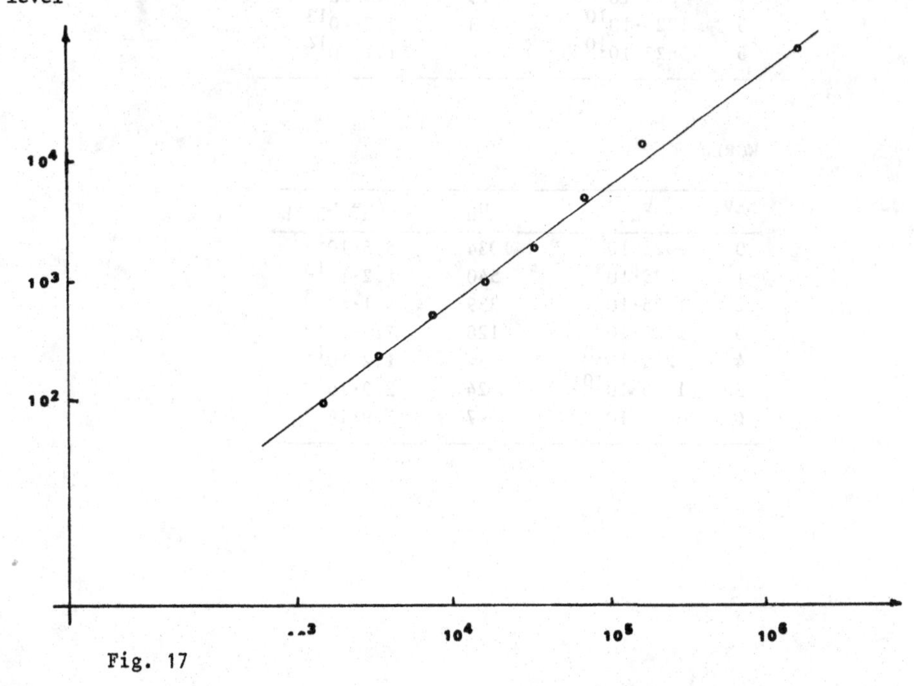


Fig. 17

From the graph we derive that
a) the average number of "local units" of a municipality belonging to a given level, is proportional to the average population of the level
b) the average number of workers (which can give us the information on the quality, on the organization and the efficiency of the service) is also proportional to the number of inhabitants of the class.

Therefore their product (which gives us the "value" of a city) is proportional to the square of the average population of the level.

We identified 10 levels, we learnt that the elements in each level are the cities, we found a criterion in order to assign a "value" to each element (the value of a municipality is proportional to the square of the average number of the population of the level).

We can conclude that the cities constitute a HMS. Applying to this system the principle of "invariance under refinement transformation" we derive the distribution law of the municipalities in the different levels. The comparison with the actual distribution of the population in different countries is very satisfying.

EUROPE 1960

| LEV. | $v_h$ | $n_h$ | $A_h = n_h v_h$ |
|---|---|---|---|
| 0 | $5625 \cdot 10^6$ | 319 | $1.8 \cdot 10^{12}$ |
| 1 | $225 \cdot 10^8$ | 186 | $4.2 \cdot 10^{12}$ |
| 2 | $1225 \cdot 10^8$ | 75 | $9.2 \cdot 10^{12}$ |
| 3 | $5625 \cdot 10^8$ | 29 | $1.6 \cdot 10^{13}$ |
| 4 | $225 \cdot 10^{10}$ | 15 | $3.4 \cdot 10^{13}$ |
| 5 | $1225 \cdot 10^{10}$ | 3 | $3.7 \cdot 10^{13}$ |
| 6 | $5625 \cdot 10^{10}$ | 2 | $1.1 \cdot 10^{14}$ |

WORLD 1960

| LEV. | $v_h$ | $n_h$ | $A_h = n_h v_h$ |
|---|---|---|---|
| 0 | $5625 \cdot 10^6$ | 1034 | $5.8 \cdot 10^{12}$ |
| 1 | $225 \cdot 10^8$ | 540 | $1.2 \cdot 10^{13}$ |
| 2 | $1225 \cdot 10^8$ | 339 | $4.1 \cdot 10^{13}$ |
| 3 | $5625 \cdot 10^8$ | 128 | $7.2 \cdot 10^{13}$ |
| 4 | $225 \cdot 10^{10}$ | 54 | $1.2 \cdot 10^{14}$ |
| 5 | $1225 \cdot 10^{10}$ | 24 | $2.9 \cdot 10^{14}$ |
| 6 | $5625 \cdot 10^{10}$ | 7 | $3.9 \cdot 10^{14}$ |

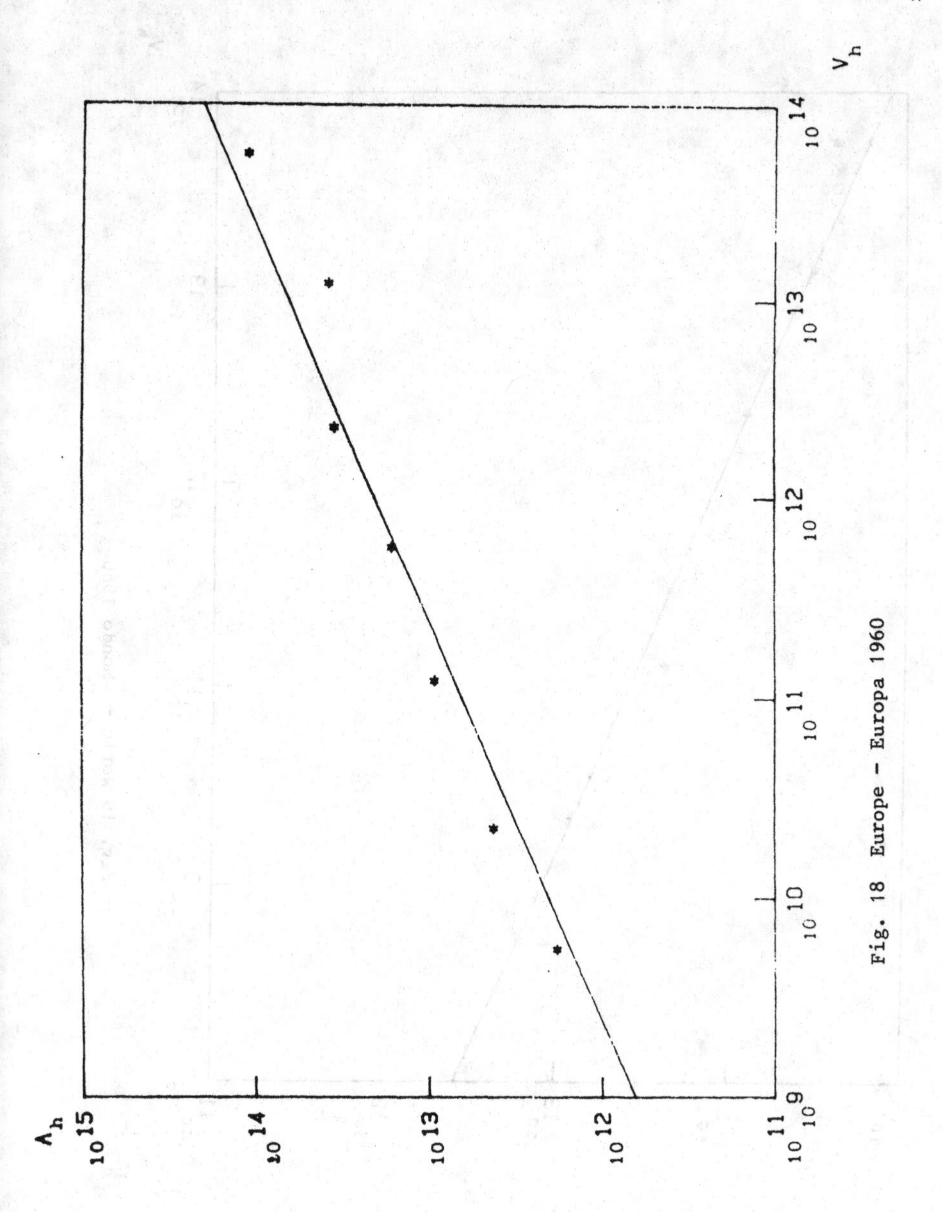


Fig. 18 Europe - Europa 1960

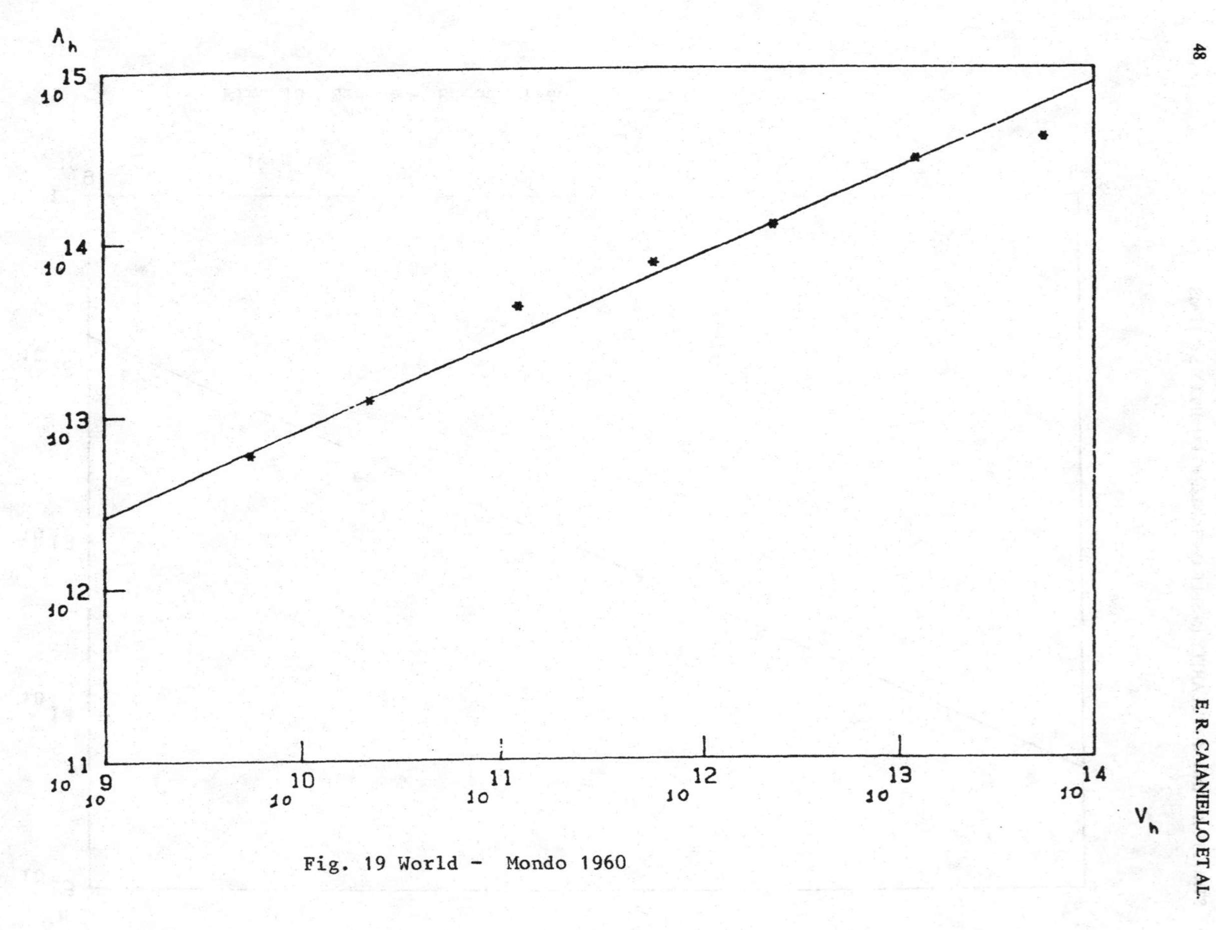


Fig. 19 World - Mondo 1960

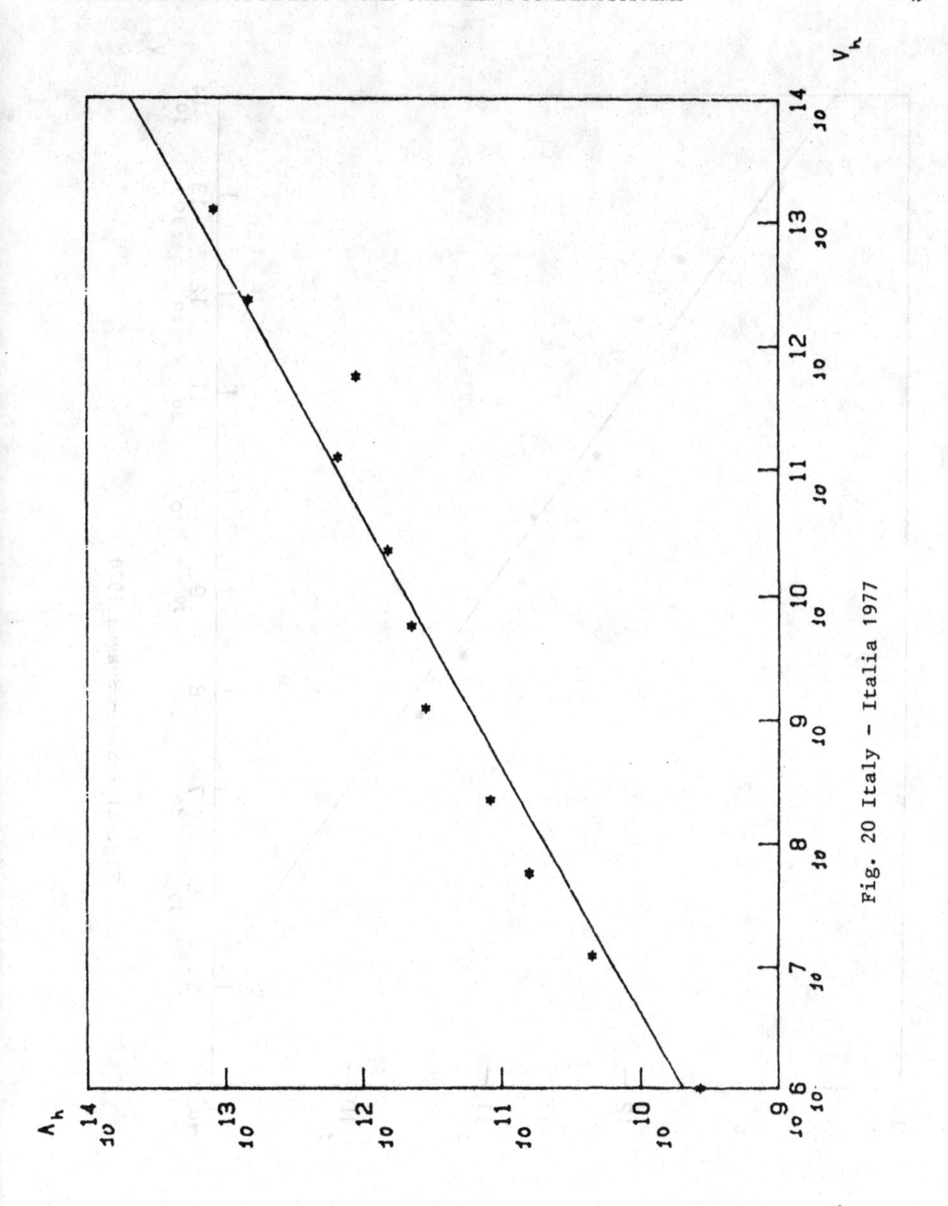


Fig. 20 Italy - Italia 1977

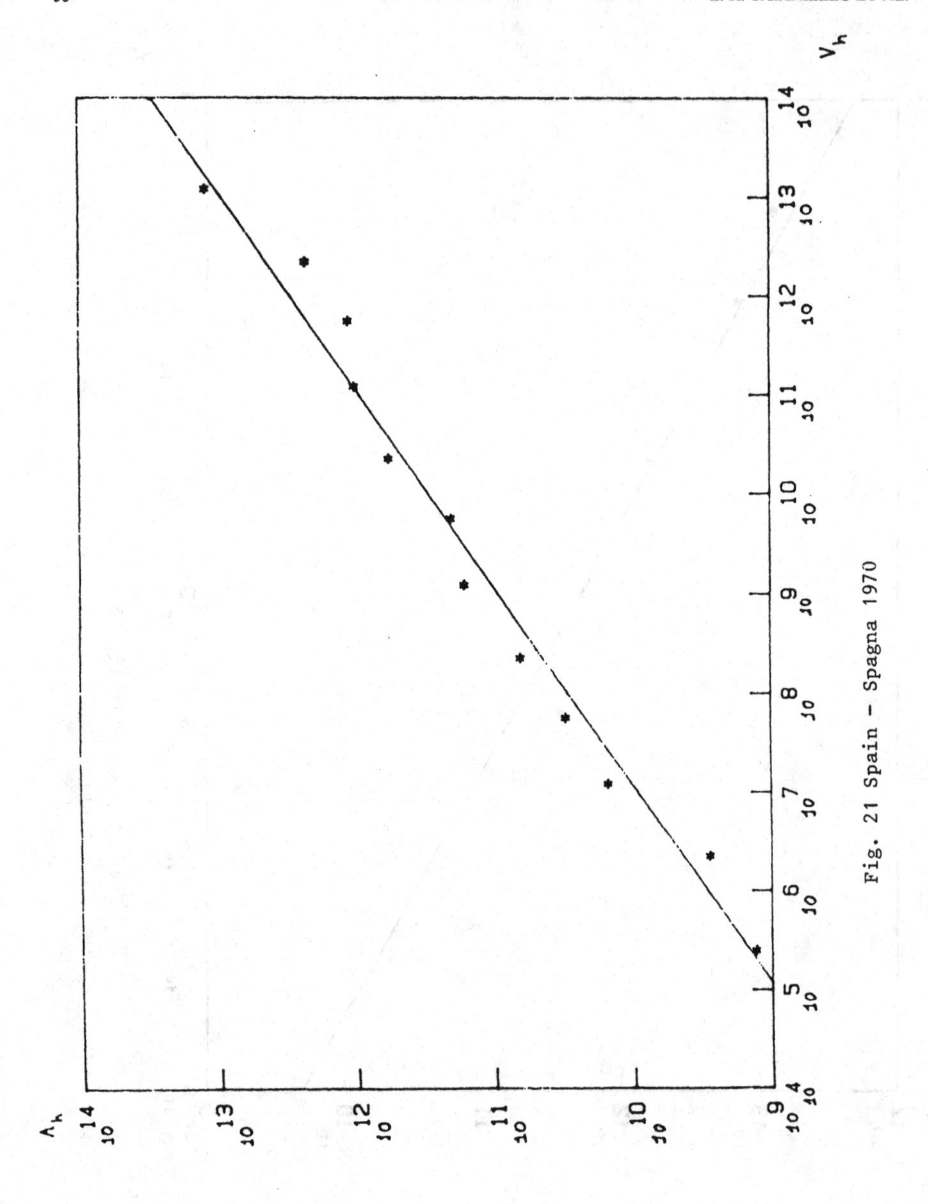


Fig. 21 Spain – Spagna 1970

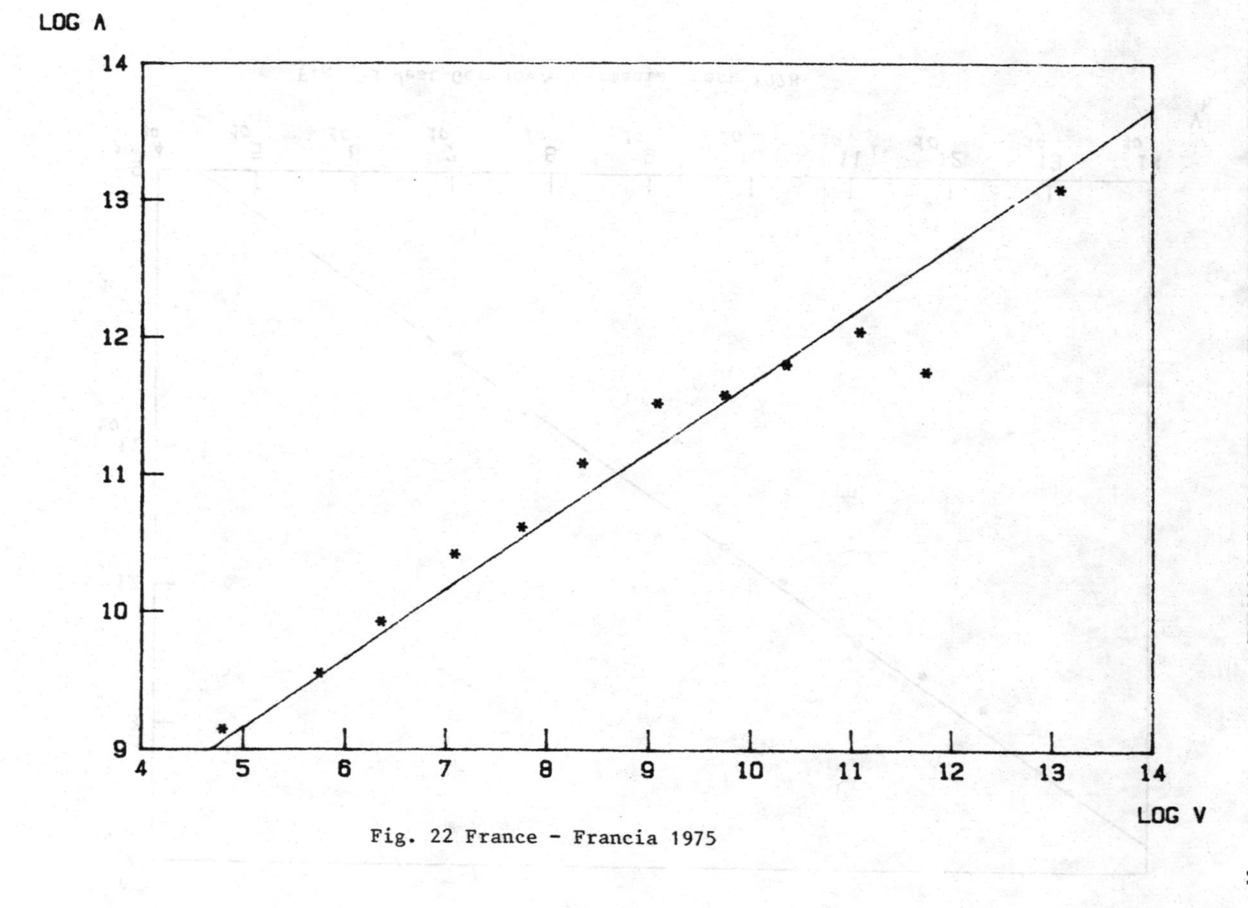


Fig. 22 France - Francia 1975

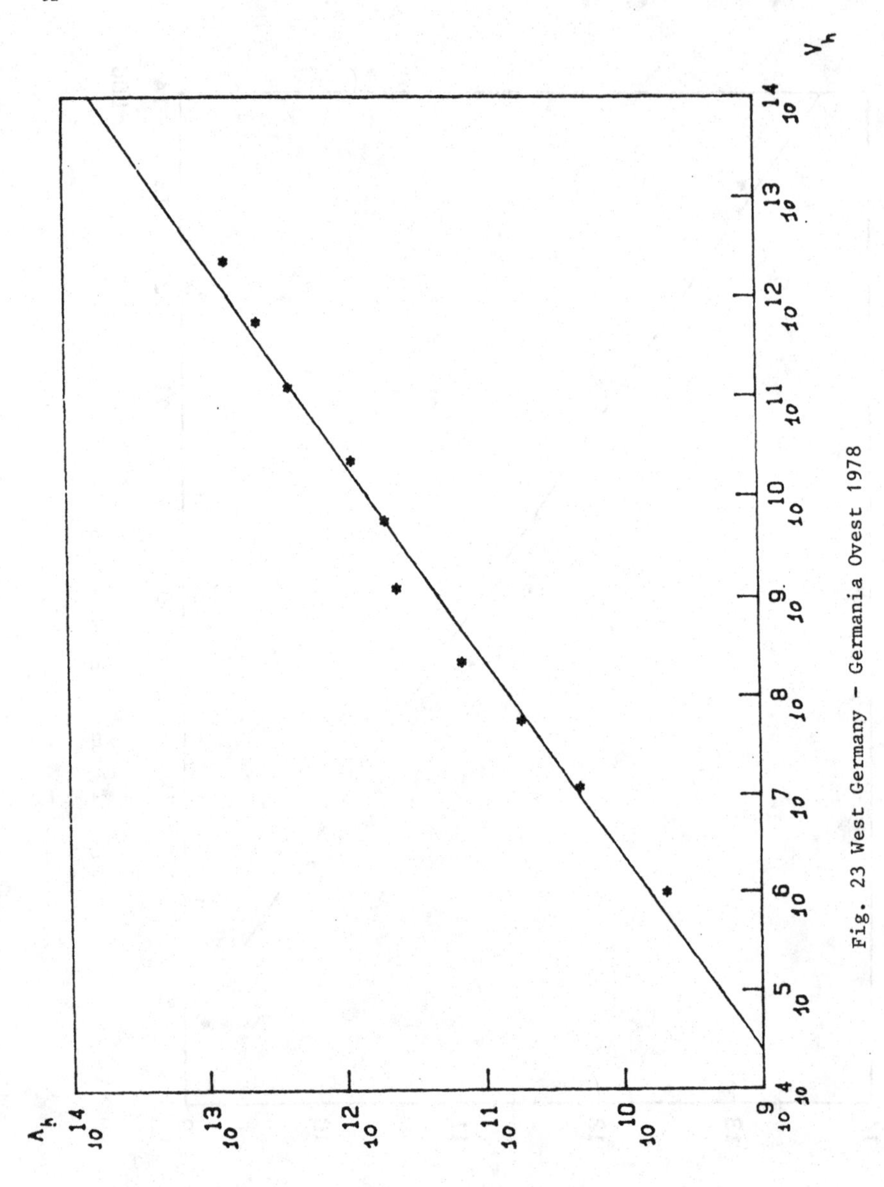


Fig. 23 West Germany - Germania Ovest 1978

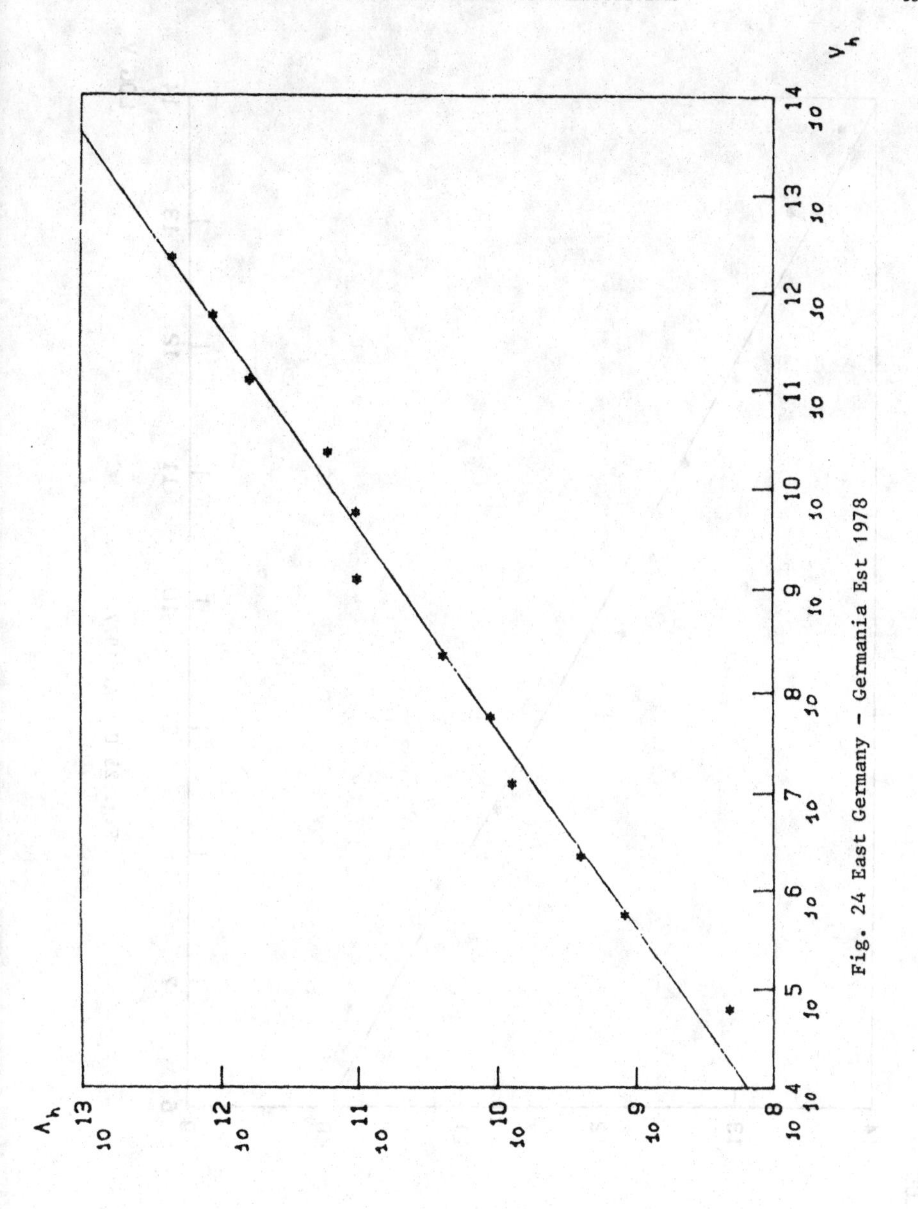


Fig. 24 East Germany - Germania Est 1978

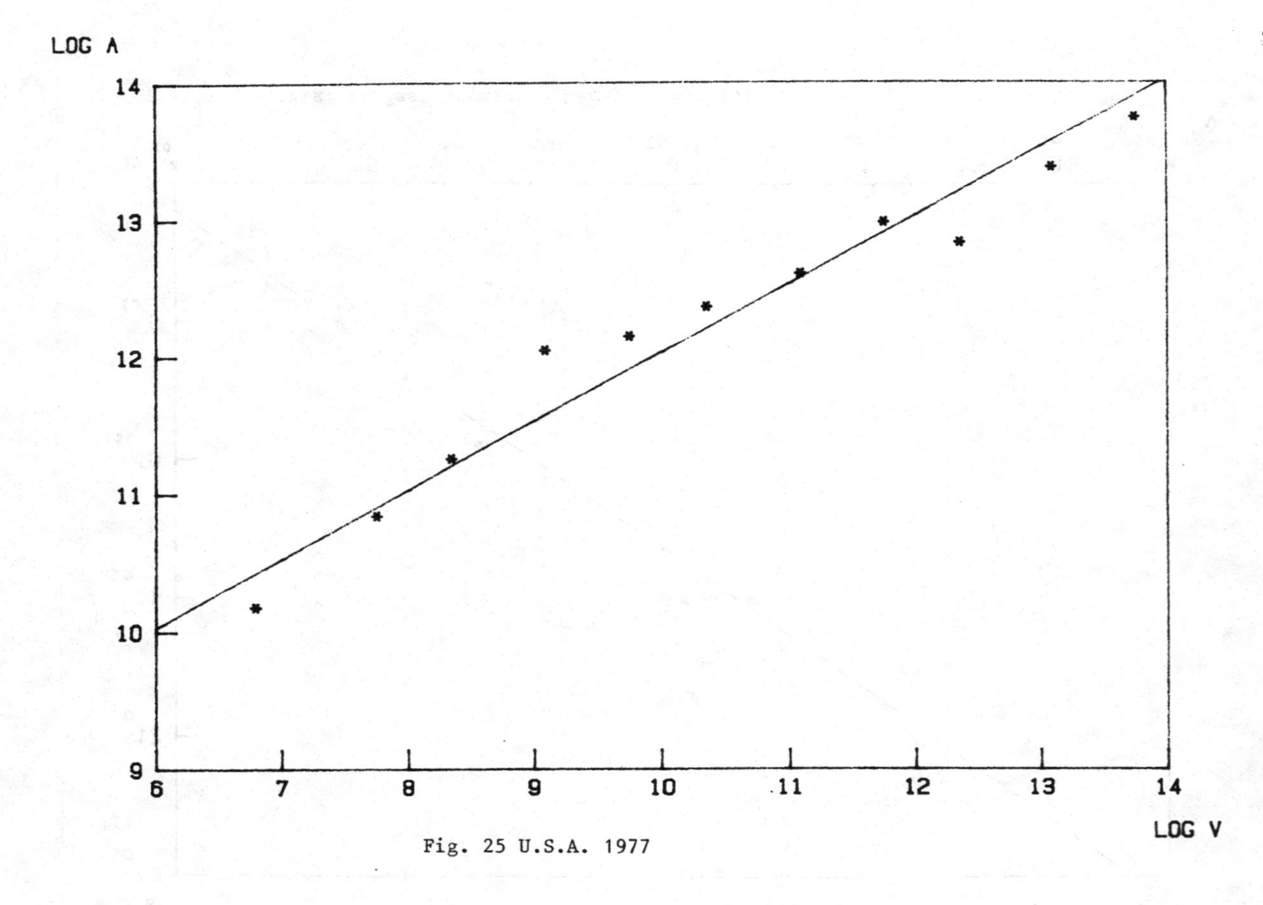


Fig. 25 U.S.A. 1977

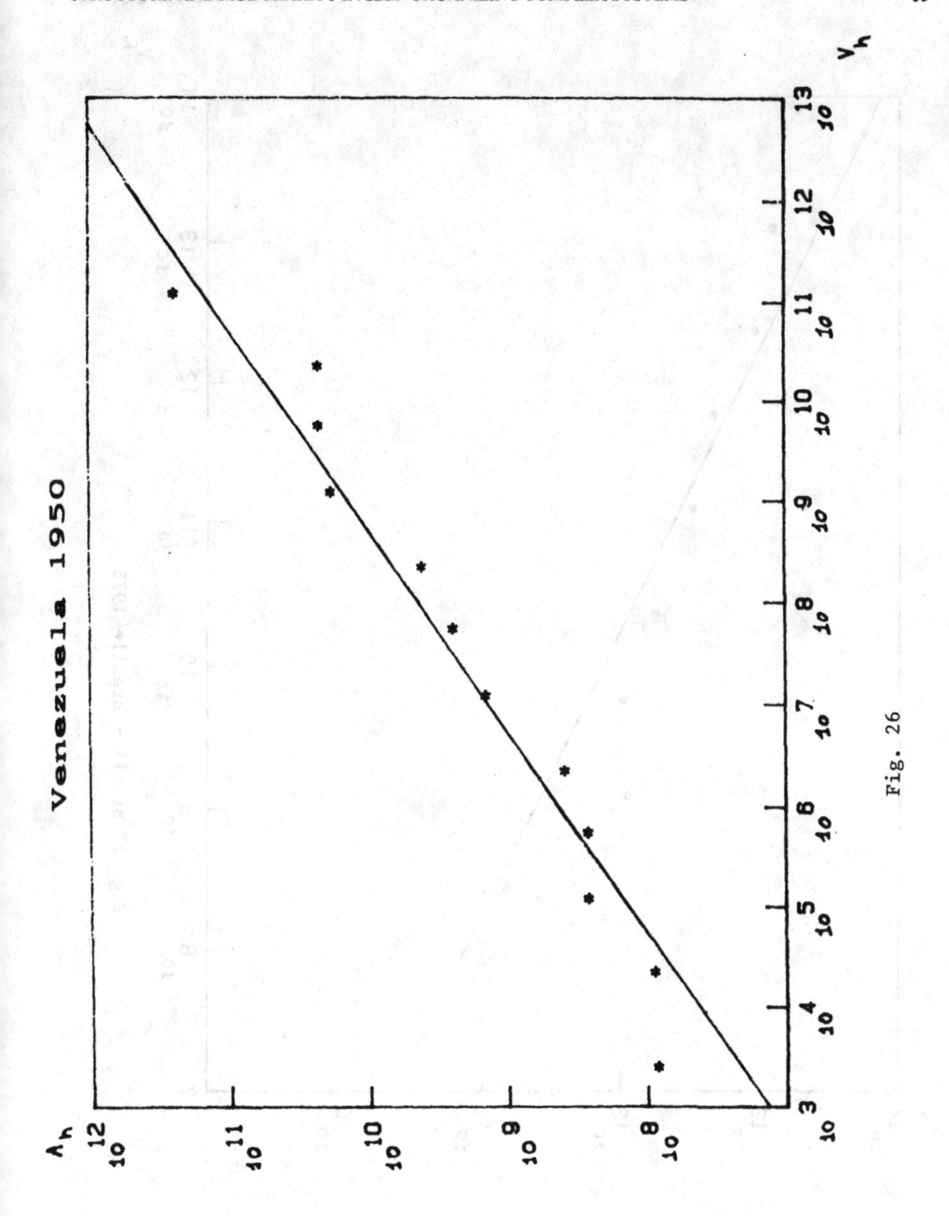
Venezuela 1950
$A_h$
$V_h$
$10^{12}$
$10^{11}$
$10^{10}$
$10^{9}$
$10^{8}$
$10^{3}$
$10^{4}$
$10^{5}$
$10^{6}$
$10^{7}$
$10^{8}$
$10^{9}$
$10^{10}$
$10^{11}$
$10^{12}$
$10^{13}$

Fig. 26

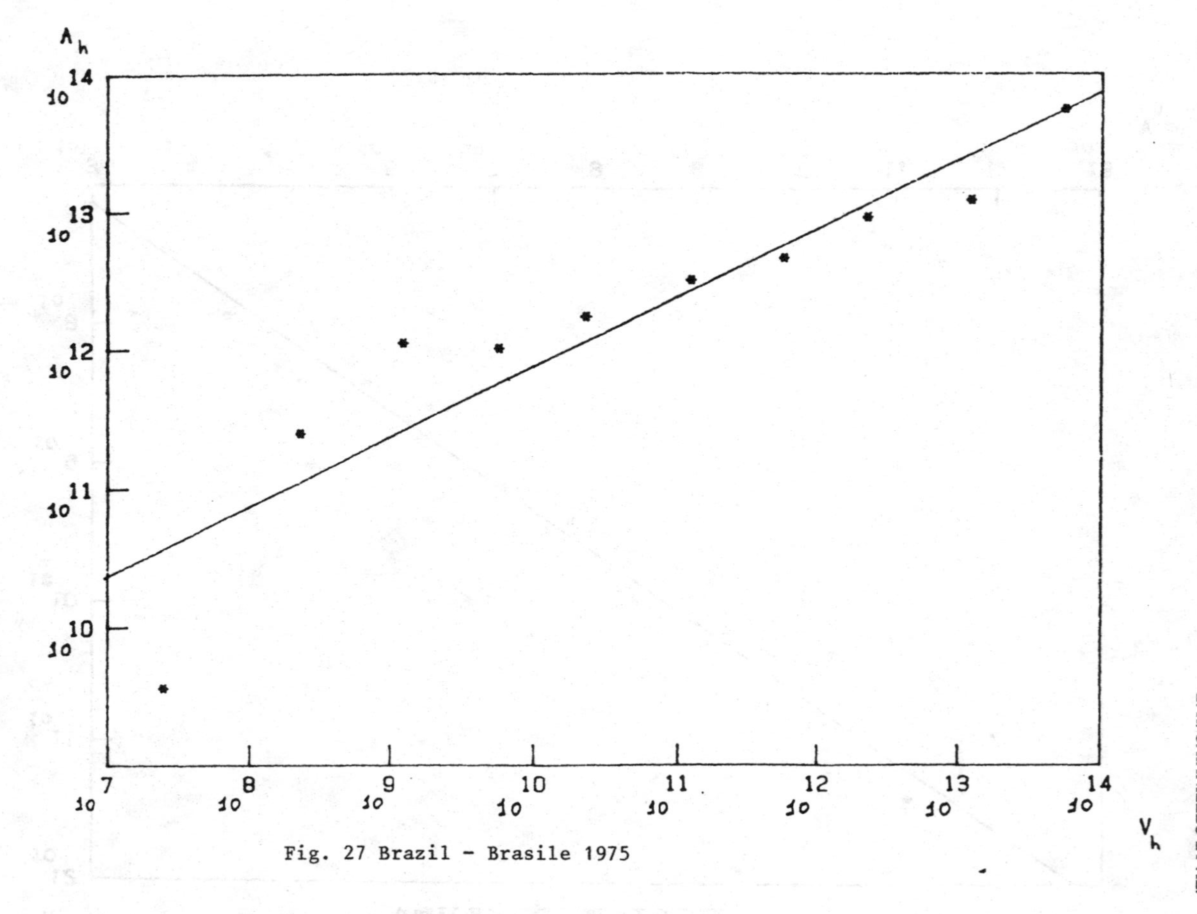


Fig. 27 Brazil - Brasile 1975

REFERENCES

1. N.S.Trubetskoj: Grundžüge der Phonologie, Praga (1939).
2. E.R.Caianiello: 'Some remarks on organizations and structures', Biol. Cybernetics, 26, 151 (1977).
3. E.Jacques: Levels of abstraction in logic and human action, Heinemann, London (1978).
   E.Jacques: A general theory of Bureaucracy, Heinemann, London (1976).
4. J.Virirakis: 'The minimization of energy as determinant of the grouping of community facilities', Ekistics, 199, 503 (1972).
   R.Carneiro: 'Scale analysis as an instrument for the study of cultural evolution', Southwestern Journal of Anthropology, 18, 149 (1962).
   R.Carneiro: 'The measurement of cultural development in the ancient near east and in anglo-saxon England', Transactions of the New York Academy of Science, 31, 1013 (1969).
5. J.Von Neumann:The computer and the brain, Yale University, New Haven (1958).
6. E.R.Caianiello, G.Scarpetta, G.Simoncelli: 'A systematic study of monetary systems', Int.Journal of General Systems, 8, 81 (1982).
   E.R.Caianiello, G.Scarpetta, G.Simoncelli: 'Sulla legge di distribuzione delle monete', Rassegna economica, 44, 771 (1980).
7. J.C.Hentsch: 'La circulation des coupures qui constituent une monnaie', Journal de la Société de Statistique de Paris, 4, (1973).
   J.C.Hentsch: 'Calcul d'un crietère qualitatif pour les séries de valeurs definissant l'échelonnement des signes monetaires', Journal de la Société de Statistique de Paris, 4, (1975).
   J.C.Hentsch: 'Distribution de la monnaie fiduciaire entre les coupures qui la représentent', Journal de la Société de Statistique de Paris 4, (1983).
8. W.Christaller: Die zentralen Orte in Süddeutschland, Jena (1933).
9. E.R.Caianiello, M.Marinaro, G.Scarpetta, G.Simoncelli: 'The population distribution as a hierarchical modular system', Preprint (1984).

# HIERARCHY AND MODULARITY IN NATURAL LANGUAGES

Alberto Negro and Roberto Tagliaferri (1)
Stefano Tagliaferri (2)

(1) Dipartimento di Informatica ed Applicazioni
Facoltà di Scienze
Università di Salerno
84100 Salerno

(2) Scuola di Perfezionamento in Scienze Cibernetiche
e Fisiche
Facoltà di Scienze
Università di Salerno
84100 Salerno

INTRODUCTION. The analysis of natural language is tackled in this work on the basis of a methodology that is the result of many years of articulate research [1,9,2]. Their origin is tied to the classic neuronic model proposed in 1961 by Caianiello [1]. The generality of such a model, in contrast to the specificity of intelligence, has stimulated a series of researches, which in a very natural way have led to the paradigm that intelligence is a form of organization [2]. This result has immediately demanded a systematic research of all natural forms of organization from which a definition of an interpretative model of complex systems can be derived: the hierarchical modular system [2,3, 4].

In this perspective, the study of natural languages seemed the perfect choice for different reasons: these systems, which have naturally developed and self-organized, are obviously "structures": the knowledge of the mother-tongue known by the experimenter is particularly detailed; the amount of material available for experimentation is enormous.

Further, the choice of the written natural language is motivated by the necessity to avoid that processes not connected with mere communication may enter the examined system.

The hierarchical modular model has the property of being sufficiently simple, but not enough to divert the system under examination from its essential aspects. In fact it has already been applied to other natural systems with great success: in an economic context to the distribution of coins; in a sociological context to the distribution of the population in a territory [3,4,7].

An interesting aspect obtained in this first analysis on languages is connected with the problem of determining the natural function of



*E. R. Caianiello and M. A. Aizerman (eds.), Topics in the General Theory of Structures, 59–68.*
*© 1987 by D. Reidel Publishing Company.*

the elements at each level; in contrast with the simple case of the monetary system, in which the value is given simply by the denomination of the coin, here the value is a more complex function; this is surely due to the specific characteristics of the system, and also to elements which for the moment are not easily measurable, such as, for instance, the semantics capacity and the presence of redundance.

## 1. LANGUAGE AS A SELF-ORGANIZING SYSTEM

The capacity to have a language is probably the principal characteristic which distinguishes mankind from other animals. It is also true that many animals have complex communication systems, but up to what actual studies have discovered, these systems have a limited capacity in transmitting information, not comparable to human language which is able to produce a very large group of concepts and meanings.

An important aspect of language is its link with the processes of human knowledge. The language of a population can be considered an archive of past life and of the experience stored up from previous generations; at the same time, it gives to the future generation an interpretation scheme of the universe. Languages can be distinguished also for other reasons, that we can trace back to the arbitrarity of the sign [5], and to language ambiguity itself. Therefore the structure of the word is a freely developing one.

All these processes give rise to self-organizing processes in language which are different between populations,in relation to factors associated with the environment in which they live and its particular structure.

However, the self-organized nature of a language may be still clearly recognizable if, instead of synchronic analysis, done by comparing the ways of adaptation of different languages in the same period of time, we proceed with a dyacronic analysis in time, through the study of the ways of evolution. Berlin and Kay [6] have compared words indicating fundamental colours in a very large spectrum of languages and cultures, coming to the conclusion that there is a link between the complexity of the system of colours and the evolution stage on the cultural scale. In this way, primitive societies, like some of the tribes which still live in the jungles of New Borneo, have adopted a very simple system, which is formed only by two terms: black and white: more advanced societies have adopted a system composed of eleven colours. The passage from one extreme to the other goes through stages following a given pattern. After black and white, red has been introduced, after it green or yellow, blue, then brown and finally only some or all of the group of colours: pink, orange, grey and violet. These stages of growing complexity correspond to different stages in the technological growth of a society. Language is subject to evolution; the consequence of this phenomenon is given to us first of all on a phonetic level. There are also consequences on the semantic power of a language. An example is given by the final 's' in the French language, which stopped being used in the fifteenth century. When this happened, it was impossible to distinguish the plural sound of many words. This situation caused a self-organizing process in the French language, so that the definite article, which until that time had been used in an optional way, was used to distinguish the singular from the plural and so its use became compulsory. The disappearance of the declension of the Latin language in the modern languages born from it is another good example of the intrinsically self-organizing nature of language. The loss of the endings was replaced by the introduction of the preposition, some of these particles being already in use before the disappearance of the declension. The important distinction between subject and complement was not realized through the differences of the endings, but through a wider syntactic stiffness; then in the modern Italian language, the subject, in affermative sentences, almost always preceeds the verb, while the object is placed after. In this way, the passage to the "neo-latin" languages has occured with the constant organization of new strategies, that counter-balance the losses caused by evolution.

## 2. THE ITALIAN LANGUAGE WRITTEN AS A HIERARCHICAL AND MODULAR SYSTEM

In this section we are going to deal with the study of language from a new point of view, in the light of the interpretative paradigm of complex systems, consisting in the hierarchical modular structure. A system is called hierarchical when we can recognize in it a structural organization on different levels; we call it modular when the value of elements on the different levels is ruled by the law:

$$V_h = M^h$$

with M being constant and $h = 0,1,...H$. A very interesting transformation applicable to these systems is the refinement [ 2 ] which is a redistribution of the elements on a greater number of levels. The invariance of the mean value of the elements with respect to refinement transformations allows the derivation of the distribution law of the elements on the levels [ 2,3,4,7 ]. When we try to recognize the hierarchical modular structure of a system, all we need to do is to solve two problems: the first is tied to the identification of the levels, the second is linked to the search for a criterion to assign the value to the elements of each level. In our case the first problem can be solved by means of two methodologies:

a) reform the grammatical syntactic structure to obtain suggestions for the identification of the levels [10 ].

b) operate more than one refining transformation on the system formed by the two extreme levels, which in general are of easy identification.

If we follow the first methodology, it is very normal to put at the zero level all the letters of the alphabet which are "atoms", that is the elementary bricks that are used to build a great number of written languages. It is well-known that the smallest relevant group of letters in a language is the syllable which we put therefore on the first level. The next level consists of words. Words put together form "predicati" which represent the elements belonging to the third level. Then we find "proposizioni", "frasi" and "periodi" which we associate respectively to the fourth, fifth and sixth levels. We have now identified seven levels which we can use to analyze the written Italian language. Another level which is not syntactic, but structural of the written language is the "capoverso", which in a written text is easily found as the group of "periodi" found in a paragraph. We have chosen a sample extracted from the Italian newspaper "Il Mattino" printed on 25th September, 1984, with a total of 14 articles of 49751 letters (including the apostrophies). The choice of articles from newspapers (one of these is shown in appendix A) was made to obtain a greater number of authors with a limited number of letters.

We have also identified for our specific sample two other levels: the titles and subtitles, and the articles. We added them as levels 8 and 9, but these are a characteristic of the sample under examination, not of the language. The addition or the cancellation of levels with respect to the eight initially identified, can also become a common fact, and this depends on the type which we analyzed: a novel, a philosophical text, an article on foreign politics, an advertisement have internal organizations which are very different from each other. The number of elements in levels 0,1,2 (corresponding respectively to letters, syllables and words) has been found automatically by computer. For the calculation of the syllables we used an algorithm which exploited the well-known property: in the Italian language every syllable contains at least one vowel. This means that at the end of the calculation, it is necessary to locate only the vowels and solve a few cases of diphtongs and triphtongs. For the higher levels we resort to semi-automatic algorithms which need human intervention. We used the following conventions:

a "predicato" is a subject or a complement with all the attributes of a "predicato verbale";

a "proposizione" is identified by the verb from which it is characterised, expressed or understood;

a "frase" is the set of "proposizioni principali" and their subordinates;

a "periodo" is a set of "frase" identified by a full-stop or by an exclamation or question mark;

a "capoverso" is a set of "periodi" identified typographically by a paragraph.

Analyzing the hierarchical structure of our sample it is possible to find two more levels: the level of titles and subtitles and the level of articles from newspapers. In table 1 we report the results of the calculation for each level.

Following the second methodology, we first identify the extreme levels, which are the letters of the alphabet and then we analyze the articles;

we calculate the number of elements which are part of two levels and we proceed with a refinement of order 9. From the experimental result shown in table 2 one can find that there is a good correspondence between the levels thus found and those obtained with the first strategy.

The second problem, that of the assignment of values to the elements in the different levels, is more complicated to solve. We assign one to the value of the letters, which are elements of the level zero; we could try to give the elements of the next levels a value equal to the average number of letters necessary to compose those elements. This type of criterion, even though it allows us to find the modularity of the hierarchical structure, takes into consideration only one aspect, the structural one, leaving out other important aspects, such as, for instance, the semantic capacity associated with the elements.

Therefore, if we have to use our model [2,3] we have

$$N_h(p) = N_0(p)\ M^{-h/2p}$$

where $h$ is the level, $p$ is the degree of the refinement. In our case $p = 9$ and having given this value $M \longleftarrow M^{1/p}$ , we have

$$N_h = N_{h-1} \sqrt{M} \qquad \text{and}$$

$$V_h = M^h\ V_0$$

$V_0$ being equal to 1 by choosing a suitable scale.

This assignment of the module leads us to take into consideration all the important aspects (semantic, syntactic or of other nature) that otherwise would be left out.

The result of the experimental analysis is reported in table 3 and fig.1, where the straight line represents the canonical distribution.

It is clear that this is only a very preliminary analysis, even though it has positive results; other results are necessary with more specific samples.

TABLE 1

| h | Nh |
|---|---|
| 0 | 49751 |
| 1 | 22215 |
| 2 | 8819 |
| 3 | 3922 |
| 4 | 1199 |
| 5 | 524 |
| 6 | 310 |
| 7 | 98 |
| 8 | 39 |
| 9 | 14 |

TABLE 2a

| h | $N_h$ |
|---|---|
| 0 | 49681 |
| 1 | 20033 |
| 2 | 8078 |
| 3 | 3257 |
| 4 | 1313 |
| 5 | 530 |
| 6 | 214 |
| 7 | 86 |
| 8 | 35 |
| 9 | 14 |

$N_h = N_H M^{H-h}$

TABLE 2b

| h | $N_h$ |
|---|---|
| 0 | 49751 |
| 1 | 20061 |
| 2 | 8089 |
| 3 | 3262 |
| 4 | 1315 |
| 5 | 530 |
| 6 | 214 |
| 7 | 86 |
| 8 | 35 |
| 9 | 14 |

$N_h = N_0 M^{-h}$

TABLE 3

| h | $N_h$ | $V_h$ | $x=\log(V_h)$ | $y=\log(N_hV_h)$ |
|---|---|---|---|---|
| 0 | 49751 | 1 | 0 | 10.81 |
| 1 | 22215 | 6 | 1.79 | 11.80 |
| 2 | 8819 | 38 | 3.69 | 12.72 |
| 3 | 3922 | 233 | 5.45 | 13.73 |
| 4 | 1199 | 1431 | 7.27 | 14.36 |
| 5 | 524 | 8801 | 9.08 | 15.34 |
| 6 | 310 | 54128 | 10.90 | 16.64 |
| 7 | 98 | 332908 | 12.72 | 17.30 |
| 8 | 39 | 2047515 | 14.53 | 18.20 |
| 9 | 14 | 12593040 | 16.35 | 18.99 |

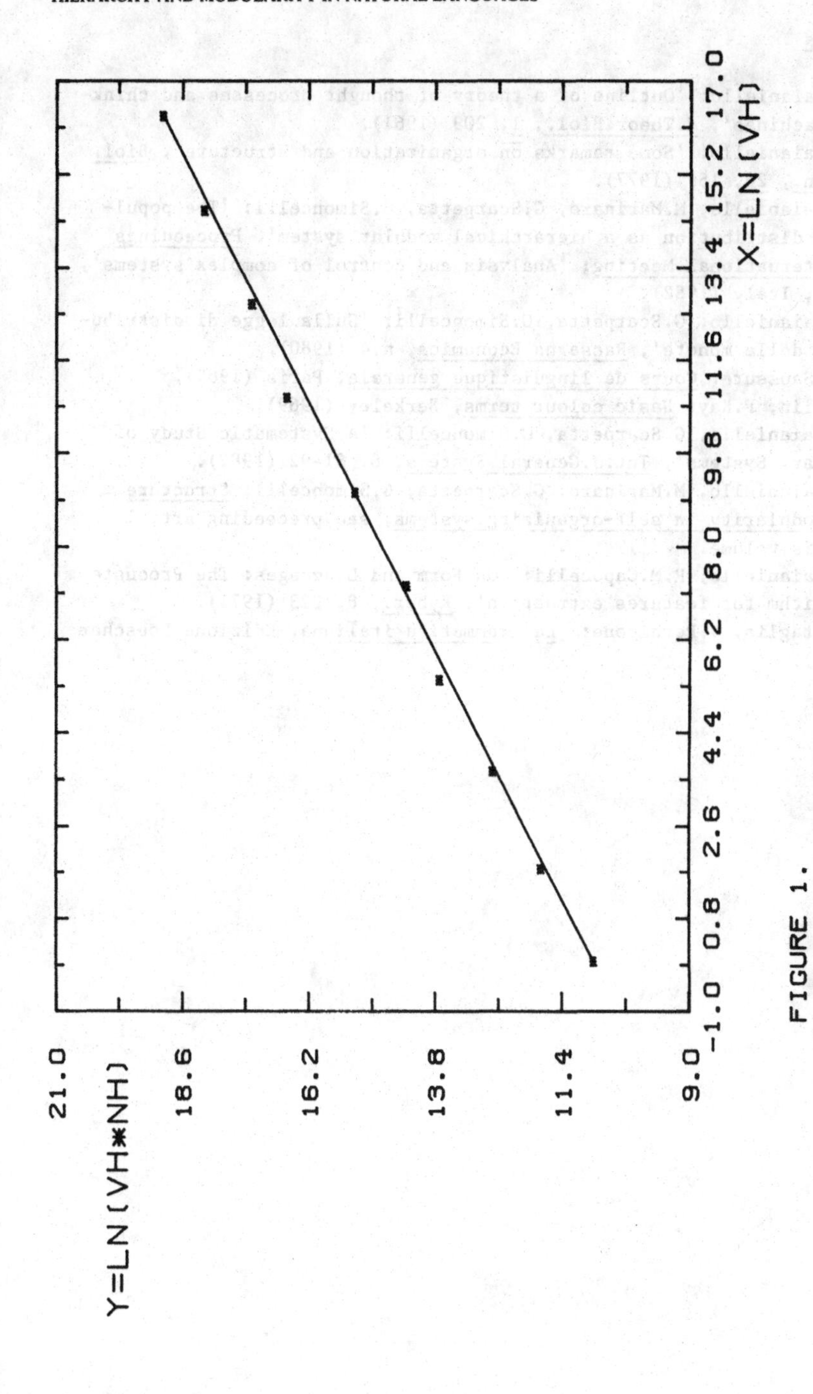


FIGURE 1.
ARTICLES FRØM IL MATTINØ

## References

1. E.R.Caianiello: 'Outline of a theory of thought processes and thinking machines', J.Theor.Biol., 1, 209 (1961).
2. E.R.Caianiello: 'Some remarks on organization and structure', Biol. Cybern., 26, 151 (1977).
3. E.R.Caianiello, M.Marinaro, G.Scarpetta, G.Simoncelli: 'The population distribution as a hierarchical modular system', Proceedings of International Meeting; 'Analysis and control of complex systems', Capri, Italy (1982).
4. E.R.Caianiello, G.Scarpetta, G.Simoncelli: 'Sulla legge di distribuzione delle monete', Rassegna Economica, n.4 (1980).
5. F.De Saussure: Cours de linguistique generale, Paris (1965).
6. B.Berlin, P.Kay: Basic colour terms, Berkeley (1969).
7. E.R.Caianiello, G.Scarpetta, G.Simoncelli: 'A Systematic Study of Monetary Systems', Int.J.General Systems, 8, 81-92 (1982).
8. E.R.Caianiello, M.Marinaro, G.Scarpetta, G.Simoncelli: Structure and modularity in self-organizing systems, see preceeding article in this volume.
9. E.R.Caianiello, R.M.Capocelli: 'On Form and Languages: The Procuste Algorithm for features extraction', Kyber., 8, 223 (1971).
10. S.Battaglia, V.Pernicone: La grammatica italiana, Edizione Loescher (1971).

## APPENDIX A

Dichiarazioni dell'amministratore delegato Paolo Benzoni. Utenti tranquilli per tutto il mille nove cento ottanta cinque. Entro il mille nove cento novanta il grand salto all'elettronica. SIP, nessun aumento. Il trenta per cento dei nuovi investimenti al Sud.

Allo stato attuale le tariffe telefoniche non subiranno alcun aumento. Almeno fino alla fine del prossimo anno. Rassicurante ed ottimista sulla lievitazione di ulteriori addebiti per l'utente, Paolo Benzoni, amministratore delegato della SIP, intervenuto ieri al convegno sugli "atteggiamenti e tendenze" nel rapporto fra la società concessionaria e la telematica, é stato più problematico sul campo dei finanziamenti. La SIP ha un fabbisogno finanziario di oltre quattromila cinque cento miliardi annui. Secondo il piano quinquennale che la società ha presentato al CIPI, gli investimenti partono da quattromila cinque cento miliardi per il mille nove cento ottanta nove. Ad una domanda specifica, Benzoni ha risposto che almeno il trenta per cento di queste risorse sarà destinato al Mezzogiorno: "Ricordiamoci che nel Sud é dislocato il quaranta per cento di tutto l'indotto delle telecomunicazioni, mentre il peso dell'utenza non supera il venti sette - ventotto." E quasi a sottoscrivere un "appeal" che ora va tanto di moda, Benzoni ha riportato in auge anche per il suo settore la cruna inevitabile attraverso la quale dovrà passare - se lo si vuole - il decollo delle regioni meridionali. La SIP sembra una delle poche aziende IRI a sapere e potere tirare la carretta, nonostante i suoi novemila cinque cento miliardi di debiti. Non la difettano managerialità, "know-how", programmazione, e tanta voglia di fare. Di recente ha tirato un sospiro di sollievo quando sono stati approvati sia l'accordo della nuova convenzione che le assegna il monopolio delle strutture e la consistenza sul mercato per gli impianti, sia il varo del piano decennale per le telecomunicazioni. Doppiato così il duplice giro di boa, la società si trova a vivere una vigilia proficua: "Nei prossimi cinque anni - annunziai Benzoni - realizzeremo il grande salto dall'elettromeccanica all'elettrònica. Entro il mille nove cento novanta un buon numero di utenti potrà trasmettere attraverso il tradizionale cavo telefonico non solo parole, ma anche dati e immagini". La telematizzazione di massa é così irreversibile e potrà consentire all'Italia di riprendere il tram dell'Occidente che sembrava perduto. Proseguendo nell'esame della politica delle telecomunicazioni, ci si imbatte così nei "nuovi servizi a valore aggiunto" sul cui mercato si gioca il "futuro delle telecomunicazioni". Al posto dei fili attuali, per dialogare con due persone o due entità a distanza l'una dall'altra oggi o subito dopo, cioé nel mille nove cento ottanta cinque, potranno farlo sotto forma digitale: sarà come inviare un trenino in direzione del nostro partner, sulla locomotiva c'é il suo indirizzo e sui vagoni i dati richiesti. Il sistema si chiama Itapac. Ma non é il solo servizio aggiunto. C'é chi punta sull'"home packing", cioé un computer a basso costo che anche da casa potrà compiere operazioni bancarie. Uno esemplificazione di questo sistema é già realizzato da aziende come la Postal Market negli acquisti per corrispondenza. Non dimentichiamo poi l'"Omega mille" che, sfruttando il videotel e il videotext, può indurre due terminali telefonici a scambiarsi ogni possibile messe di dati e di immagini. Ecco un esempio: l'Italgiure mette a disposizione di un terminale tutte le sentenze della corte di Cassizione. Attraverso questa "rete fonia dati" si può approdare ad un Servizio che permette all'abbonato di essere chiamato da tutta l'Italia con un unico numero e di vedersi anche addebitate la chiamata. Vanno alacremente avanti gli studi per identificare chi si diverte a fare telefonate anonime. Questo problema era di difficile soluzione con l'elettromeccanica ancora vigente - revela Benzoni - ma una volta sfondata la soglia dell'elettronica, la sarà risolta agevolmente. Numerosi dei sistemi menzionati sono già in funzione e non c'é città del Sud di un certo peso che non sia provvista. "Oggi l'Italia é un bivio: produrre nuovi disoccupati attraverso una economica assistita o qualificare la spesa pubblica, come avviene altrove, investendo nei settori di punta come la telecomunicazioni?", si chiede l'amministratore delegato della SIP. Tutto lascia supporre che anche il nostro paese intende finalmente rimettersi in riga, seppure in ritardo, alla stregua degli altri.

DYNAMIC APPROACH TO ANALYSIS OF STRUCTURES DESCRIBED BY GRAPHS (FOUNDATIONS OF GRAPH-DYNAMICS)

M.A.Aizerman, L.A.Gusev, S.V.Petrov, I.M.Smirnova, L.A.Tenenbaum

Institute of Control Sciences, Moscov, USSR

## 1. DYNAMICS OF TREES WITH NUMBERED VERTICES

### 1.1 Introduction

In the application and analysis of dynamical problems described, for instance, by differential or difference equations, unknown functions are usually represented by "numerical" functions, which take their values on the whole real axis or on a limited subset of it. In this case the solution of the dynamical problem is a sequence of numbers variable with time, and the aim of the dynamical analysis consists either of the formation of these sequences (solutions of the problem) once initial conditions have been given, or of the search for their general characteristics: asymptotic behaviour, presence or absence of particular solutions (equilibrium conditions, periodical solutions, etc.).

Besides dynamical problems of this kind, in the applications other problems arise, where it is not an intrinsically important numerical function of the system that varies with time, but its same structure. Several examples of these kinds of problems can be given.

Example 1.1 Administrative structure. An administrative structure can often be described with a tree graph, whose vertices form the elements of the structure, whereas its sides represent the mutual subordination of the elements. If the administrative structure "lives" according to determined internal laws, or if it is deliberately changeable in such a way as to create new structures, then, step by step, an evolution process of the tree graph describing the structure will occur with time. According to the special characteristics of the "trajectory" followed by the structure, cases are possible where the dynamical process leads to the appearance of a new structure, which afterwards remains fixed, or, in other cases, because of the conditions of the dynamical process,after a series of changes, the structure may return to that of the old original type, so that we have cyclical trajectories. During the life of the administrative structure other dynamical processes are also possible.



*E. R. Caianiello and M. A. Aizerman (eds.), Topics in the General Theory of Structures, 69–136.*
*© 1987 by D. Reidel Publishing Company.*

Example 1.2 Organization of communication and service systems. As a typical example, we consider the organization of post offices, operating in a certain geographical area. The basis of the system is given by the lowest level offices, which are controlled by the respective group or area directions, which, in their turn, depend on departmental direction, and so on up to the highest level represented by the Ministry of Post and Telecommunication services. The creation of industrial or agricultural settlements or of other factors leading to the migration of the population to new territories, causes a change in the layout of the lowest element of the structure (post office); and because of the laws which regulate the dependence of the lowest level offices from the superior ones, the result is an inevitable gradual re-organization of the higher levels of the elements in the structure.

Example 1.3 Listing the parts of a specific class of objects. Let us consider, for instance, the list of elements necessary for the construction of a radio-set. It includes small parts, which can be bought independently (transistors, tubes, strengths, capacitances), and pre-made parts as well (scanning and feeding units, or amplifiers AF, etc.).These parts are then put together to form more complex pieces which appear in the producer's list and can be bought for various purposes. The complete list of the small and ready-made parts, and of the final prices forms the list of the elements of a radio-set, and these can be represented by an oriented graph, whose vertices are the elements appearing in the list, and whose sides represent the connection of some elements with others. In general such a graph is not necessarily a tree. The introduction of new kinds of components or new pieces (for instance, transistors, tape-recorders, etc.) which are added to the list, does not only cause the change of a vertex in a graph: the whole graph is reordered, because now the various pieces are constructed with new components.

Naturally, this example refers not only to the components of a radio-set, but to the listing of other types of manufactured articles, for which a new range of products and spare parts are produced.

Example 1.4 Arrangement of the associative memory of a computer. When the memory of a computer is constructed according to the associative principle, words are written in different bins of the memory and are "assembled" in units following a certain criterion; in turn, these units are arranged together according to other criteria, which are in general rougher, etc. This hierarchical structure lives and changes continuously, because, while the computer is working, not only the number of bins in the given units, or the number of units in some "sequence" changes, but the units automatically appear and disappear according to the commands. Thus, while the "instantaneous" memory state is described by a graph (in particular a tree), during the operation of the computer, we obtain a sequence of graphs.

In all these examples, the notion of "structure" is adequately represented by an oriented graph; and besides, we know intuitively that in some cases it is sufficient to consider only a relatively simple class of graphs, that of trees (or forests, whose components are trees). When real objects can be represented by graphs, static problems are generally tackled and solved. This means that the graph itself is considered fixed, while its properties are being investigated. Sometimes, two graphs are assigned and we tackle the problem of deriving a third graph from them (graph algebra). When dynamical problems are proposed for graphs, they are usually dealt with as dynamical phenomena arising during the movement along the graph; the graph itself is however considered fixed.

In examples similar to the previous ones, problems of a different kind can arise: the graph is, so to speak, an "instantaneous reproduction" of the phenomenon at a moment, and the development of the events with time (dynamic properties) is not connected with the movement along the graph, but with the variation of the graph itself. In such dynamic problems the object under investigation, that is the variable whose change with time describes the dynamic process, is the graph as a whole. With the term "graph dynamics", we label all the different methods used to describe and study such systems.

Let $x(t)$ be the graph in the instant $t$, and assume that the variation law in time of the graph can be described by a recursive process

$$(1.1) \quad x(t+1) = F\left[x(t)\right]$$

where $F$ is the operation on the graph, transforming the observed graph, at time $t$, into the graph appearing at time $t + 1$. Leaving aside for the moment the question of how this operation can be formalized, we will start by introducing the basic notation and notions which will be used for the construction of our graph dynamics. We will call $x(0)$ the initial graph, and graph-trajectory, the sequence of $x(t)$-graphs obtained from (1.1). Graph $x^*$ will be in equilibrium if we have

$$(1.2) \quad x^* = F(x^*)$$

The set of $x(0)$-graphs, for which a given equilibrium $x^*$ is established in a certain time due to (1.1), will be called attraction field or field convergent to equilibrium.

When, as an effect of process (1.1) - starting from graph $x(0)$ - we find that there is a time $t = T$ and an integer positive number $K > 1$, such that

$$(1.3) \quad x(T + K) = x(T) \; .$$

we will speak of a cycle of length $K$ (which is not to be confused with a cycle on the graph). The graph trajectory may converge to a cycle, and in this case the cycle has a certain convergence domain. Of course, it is possible that there are graph trajectories convergent neither to equilibrium, nor to a cycle; for example, when under the "evaluation rules" the number of vertices of a graph continuously increases and the graph changes without repetitions.

Similarly, into graph dynamics other notions can be introduced, which are used in "ordinary" dynamics: in particular the notion of "distance" between graphs.

This distance may reflect the intuitive idea of nearness or similarity between two graphs. For example, we can assign to the graph a scalar quantity, interpreted as an individual evaluation of the complexity or of some other characteristic of the graph; therefore, we can define the distance as the difference between two scalar evaluators. Whenever the notion of distance is used, it helps us to introduce the idea of stability of graph trajectories with respect to small perturbances, and the idea of monotonicity of processes.

In short, all the notions considered when analysing "ordinary dynamical" systems, can theoretically be used in "graph dynamics". The fact that the graph can be characterized by some "numerical characteristics" or by a set of numbers, for instance by the adjacency matrix of the graph, permits the application of usual methods of description of dynamics of systems (difference equations) to graph dynamics. All this occurs on condition that the "numerical characteristic" or the "unknown function" allows the univocal reconstruction of the graph. By following this methodology, two difficulties generally arise: firstly, usual operations on "numerical functions"characterizing the graph are not included in the class of the graph in question; secondly, graph dynamics may involve problems particular to themselves, which are not conveniently analagous to the dynamics of normal objects. for instance, the problem may arise of isolating a subgraph from a graph, which remains unchanged (or it changes "very little") along the graph trajectory; or the "collective preserving" problem consisting in the isolation of a group of vertices (a "collective") that is always subject to the same "leading" vertex along the graph trajectory.

In connection with all this, a language is needed which is helpful to represent graphs for concrete graph dynamics problems, and another one suitable for the description of the relative operators.

The first results 1,2 gained in this field will be summarized. In the next section we will introduce a subordination function with integer values, operations with it and their meaning. In section 1.3, we will build the graph dynamics equations, which allow the description of graph trajectories with time.

Chapter II enlarges on the ideas contained in the first chapter, applied to larger classes of graphs.

Finally, in chapter III, we will discuss new results: we will study direct graphs of an arbitrary nature, using graph grammar.

## 1.2 "Subordination" Functions and Operations with them

In this section we will study oriented graphs (in particular trees) whose vertices are distinct from one another. Generally, we assume that vertices are numbered: 1, 2, 3,....N. We limit our studies to graphs of a particular type, i.e. acyclic graphs and those whose correspondence is specified one-to-one . This means that our discussion concerns oriented trees and separated graphs, in which the components are oriented trees ("forests").

As for the description of graphs with numbered vertices, we will use functions with integer values; it is natural to apply the mathematical apparatus developed for them, and to translate graph dynamics problems into the language of recursion relations of functions with integer values.

When applied to graph dynamics, these relations satisfy special conditions, which are not generally met, and therefore we will have to deal, on one hand, with recursion problems of a particular kind, and on the other, with the graphic discussion of problems.

Subordination function

Consider now the class of graphs characterized by the two following conditions:

1) Each graph of the class is a tree (with a root), or it consists of several trees ("forests").
2) Unless otherwise specified, it is assumed that the number of vertices is finite and equal to N, and that the vertices of a graph are numbered with the positive integers 1,2,3,...N.

As to numeration, we impose the following condition: the number assigned to a vertex is always larger than the number assigned to the vertex to which the former is subordinated. In this way, the numeration establishes an order between the vertices of a given level of the hierarchy, which is not influenced by the order established in a higher level.

In a tree graph, we will use the following rule for the numeration of vertices: number 1 is always assigned to the root, then, the vertices directly connected with the root are numbered in a left to right order, then, in the same manner, the vertices of successive levels, etc. (see example in Fig.1.1)

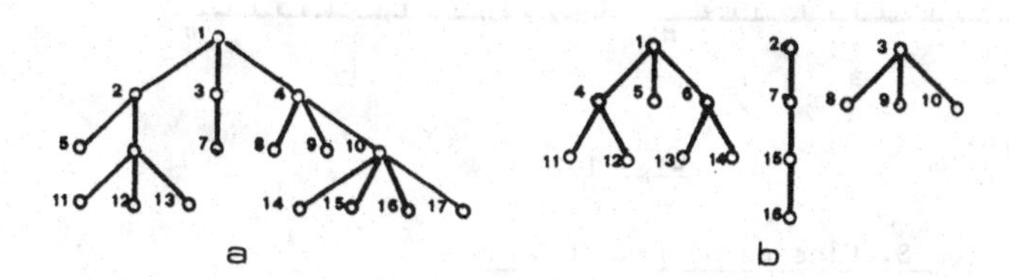


Fig. 1.1

If the graph is a forest, the numeration of its vertices occurs with a similar rule, from left to right and from above to below (see example in Fig. 1.1b).

The function $\varphi(n)$ with integer values is defined as follows: let n be the number given to a vertex; $\varphi(n)$ is defined as the number of vertices of the superior level, with which the vertex is connected. We will name this function with integer values, subordination function, or, more shortly, function S.

Although the argument assumes only integer values larger than zero, 1,2,3,..., the function $\varphi(n)$ can be equal to zero; this is useful for completing the definition of $\varphi(n)$ is cases where the vertex n is not subordinated to any other; in such a case, we will impose $\varphi(n) = 0$.

We will assume that there are N vertices in the graph, and that all the numbers between 1 and N are assigned to the vertices. The function S with integer values obviously satisfies the following conditions:

1° $\varphi(n) < n$, i.e. the number assigned to a vertex is always larger than the number assigned to the vertex to which it is subordinated.

2° $\varphi(n)$ is defined on the set of all integers from 1 to N.

Each regularly numbered tree generates a unique function S.Now, take an arbitrary function with integer values $\varphi(n)$, satisfying conditions 1° and 2°, and defined by a set of numbers 1,2,3,...N. Then, from the given function and N, we can univocally reconstruct the corresponding graph (not necessarily simply connected) with numbered vertices.

In Fig.1.2, there are examples of function $\varphi(n)$, satisfying conditions 1° and 2°, whose corresponding graphs are shown in Fig.1.1.

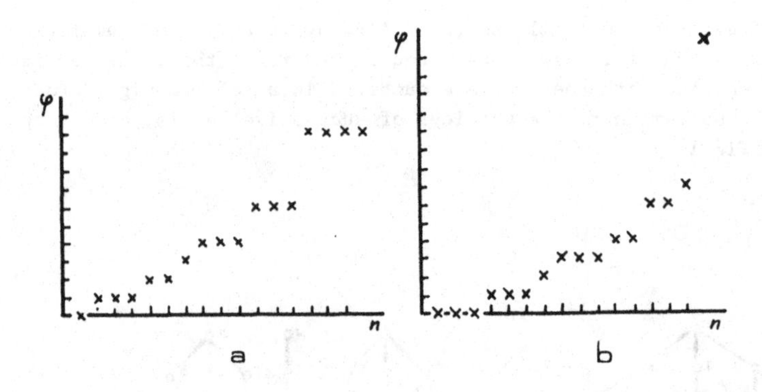


Fig. 1.2

<u>Graphs of function S. Classes of functions S</u>

On the plane n, $\varphi(n)$ (Fig.1.3) we consider the domain limited by the horizontal axis and by the bisectrix of the first quadrant (sketched region in Fig.1.3).

As a consequence of the numeration method introduced and of condition 1°, the function $\varphi(n)$ is represented by discrete points with integer coordinates, lying in a field G (including its neighbourhoods). An example of such a function is shown in Fig.1.3. The arrangement of these points within the domain is not constrained in any way, and the fact that a constraint is imposed means that a specific class of function S is being isolated.

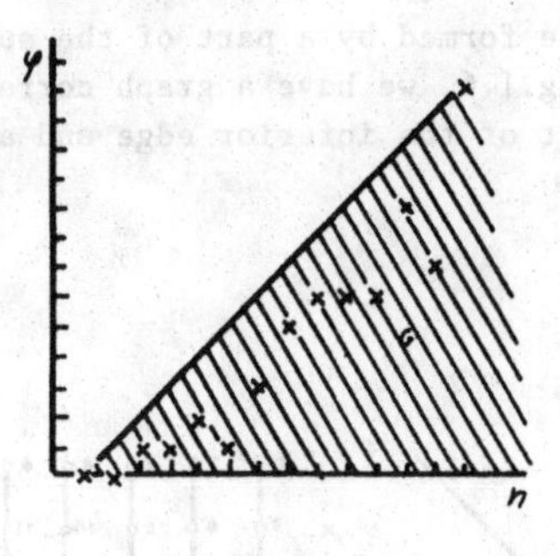


Fig. 1.3

In correspondence to functions assuming values only on the edge of domain G, we have graphs with well-defined structure: (n) = 0 corresponds to a graph which consists of a set of non-connected vertices; a function S taking values on the bisectrix, corresponds to a chain graph (Fig. 1.4).

Fig. 1.4

A graph whose structure is a "chain with a fan at the end" (Fig. 1.5)

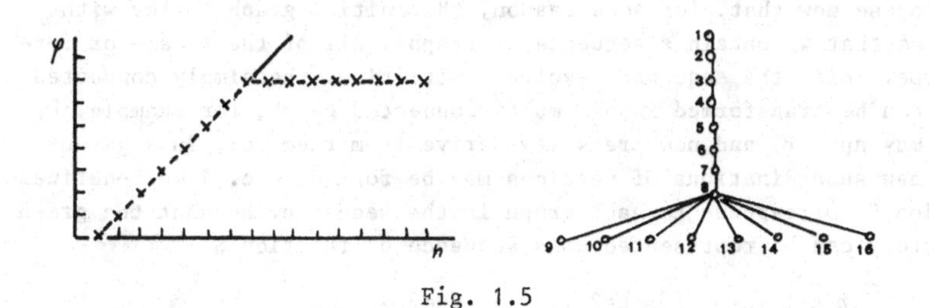


Fig. 1.5

corresponds to a broken line formed by a part of the superior edge and a horizontal segment; in Fig.1.6, we have a graph corresponding to a broken line formed by a part of the inferior edge and a segment parallel to the bisectrix.

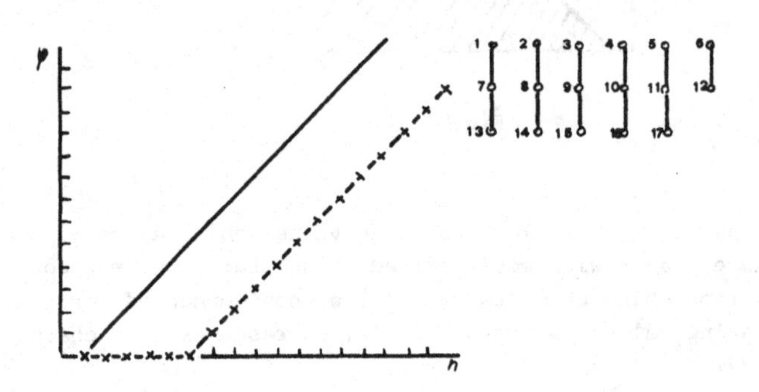


Fig. 1.6

Vice versa, in correspondence with a horizontal line $\varphi(n) = 1$, $n = 2,3,...N$, we have a "fan graph" (Fig.1.7a), and in correspondence with a broken line, obtained from a horizontal segment and from a segment inclined at 45°, we have a "fan with subordinated chains". (Fig.1.7b)

If we draw a curve continuing into domain G, and decide to impose the nearest integer number below the curve as value of $\varphi(n)$ for any n, then, a particular function S will correspond to each curve, and from this function S a particular graph. It is easily demonstrated that we have, in correspondence with the inclined line $y = \frac{1}{2}x$, a binary tree (Fig.1.7c) and in correspondence with the curve $y = \sqrt{x}$, a graph with an increasing number of subordinated vertices, as shown in Fig.1.7d.

Unary operations

So far, we have considered a single graph and the function S specifying it. Suppose now that, for some reason, the initial graph varies with time, so that we obtain a sequence of graphs, all of the tree - or forest type. While the sequence "evolves" with time, the simply connected graph can be transformed into a multi-connected graph; for example, new roots may appear, and new trees may derive from them, or, in a given tree, new subordinations of vertices may be formed, etc. A well-defined function S corresponds to each graph in the sequence, so that the graph trajectory can be represented as a sequence of function S

$$\varphi^{t=1}(n) \longrightarrow \varphi^{t=2}(n) \longrightarrow \varphi^{t=3}(n) \longrightarrow \ldots\ldots\ldots$$

Let us now consider the recurrent relation

$$(1.4) \qquad \varphi^{t+1}(n) = F(\varphi^{t}(n)) \quad ,$$

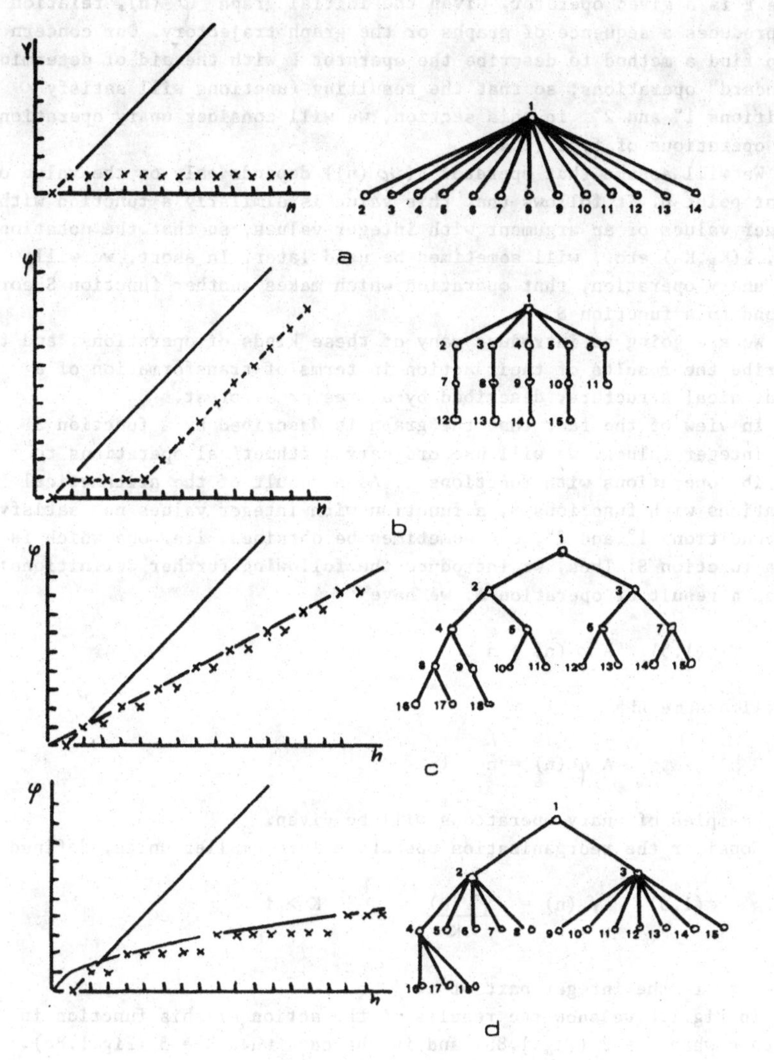


Fig. 1.7

where F is a given operator. Given the initial graph $\varphi^{\circ}(n)$, relation 1.4 produces a sequence of graphs or the graph trajectory. Our concern is to find a method to describe the operator F with the aid of determined "standard" operations; so that the resulting functions will satisfy conditions 1° and 2°. In this section, we will consider unary operations, i.e. operations of type (1.4).

We will assume that operator $F(\varphi(n))$ depends only on the value of $\varphi$ at point n. It follows that this value is similarly a function with integer values of an argument with integer values, so that the notation $F(K)$, $F(K_1,K_2)$ etc., will sometimes be used later. In short, we will call unary operation, that operation which makes another function S correspond to a function S.

We are going to introduce many of these kinds of operations, and to describe the results of their action in terms of transformation of a hierarchical structure, described by a tree or a forest.

In view of the fact that the graph is described by a function S with integer values, we will use ordinary arithmetical operations to describe operations with functions S. As a result of the arithmetical operations with functions S, a function with integer values not satisfying conditions 1° and 2°, can sometimes be obtained, i.e. one which is not a function S. Thus, we introduce the following further definitions: if, as a result of operation A, we have

$$(1.5) \qquad A\varphi(n) > n$$

we will assume that

$$A\varphi(n) = n - 1$$

Some examples of unary operations will be given.

Consider the reorganization operation into smaller units, defined as

$$(1.6) \qquad \Psi(n) = \frac{\varphi(n)}{K} \qquad K > 1$$

where a is the integer part of a.

In Fig.1.8 we show the results of the action of this function in the case when K = 2 (Fig.1.8b) and in the case when K = 3 (Fig.1.8c), applied to the three initial graphs shown in Fig.1.8a. The analysis of this example and the elementary properties of operation 1.6 reveals what approximately results from this operation: it isolates as "independent" a certain number of "leading vertices" placed at the high levels of the initial hierarchy, and so it rearranges the remaining vertices, subordinating them to "the leaders" in a more or less uniform manner.

The operation "inverse" to 1.6, i.e. that of "reorganizing into larger units", is defined as

$$(1.7) \qquad \Psi(n) = K\varphi(n) \quad , \qquad K = 2,3,\ldots.$$

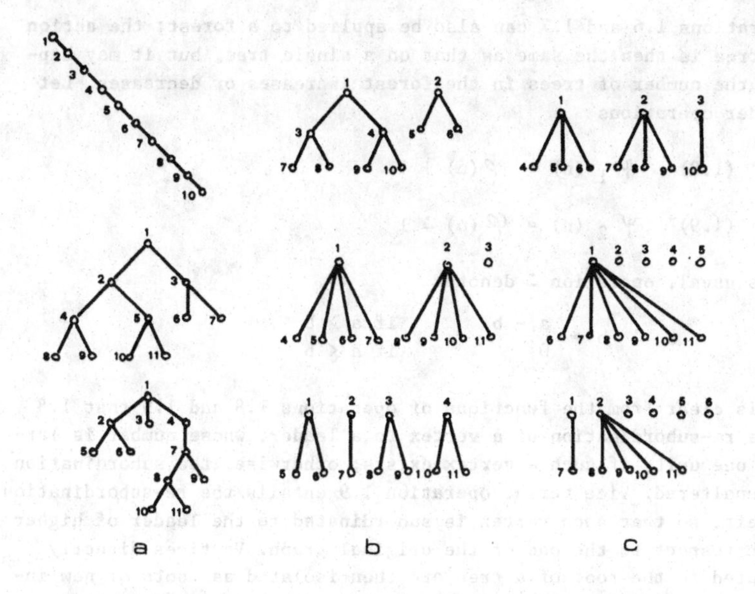


Fig. 1.8

In a certain sense, this operation acts as a stretching along the vertical axis; its action is shown in Fig.1.9, in the case when K = 2 (Fig.1.9b) and K = 3 (Fig.1.9c) for the two initial trees shown in Fig.1.9a.

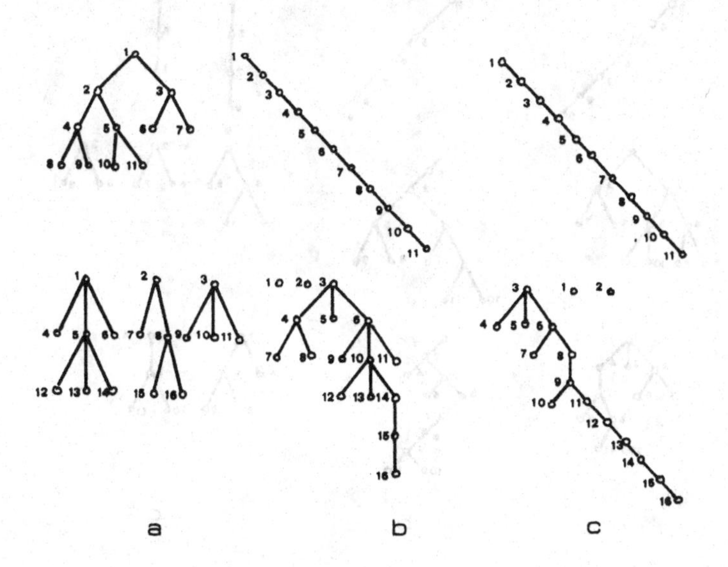


Fig. 1.9

Operations 1.6 and 1.7 can also be applied to a forest; the action on each tree is then the same as that on a single tree, but it may happen that the number of trees in the forest increases or decreases. Let us consider operations

$$(1.8) \qquad \psi_1(n) = \varphi(n) + 1$$

$$(1.9) \qquad \psi_2(n) = \varphi(n) \dot{-} 1$$

where, as usual, operation $\dot{-}$ denotes

$$a \dot{-} b = \begin{cases} a - b & \text{if } a > b \\ 0 & \text{if } a \leqslant b \end{cases}$$

It is clear from the functions of operations 1.8 and 1.9 that 1.8 shows the re-subordination of a vertex to a leader, whose number is larger that one unit, if such a vertex exists; otherwise, the subordination remains unaltered. Vice versa, operation 1.9 entails the re-subordination to the left, so that each vertex is subordinated to the leader of higher rank with respect to the one of the original graph. Vertices directly subordinated to the root of a tree are then isolated as roots of new individual trees. Fig.1.10a shows the initial trees, Fig.1.10b the action on them of operation 1.8, Fig.1.10c the action of operation 1.9.

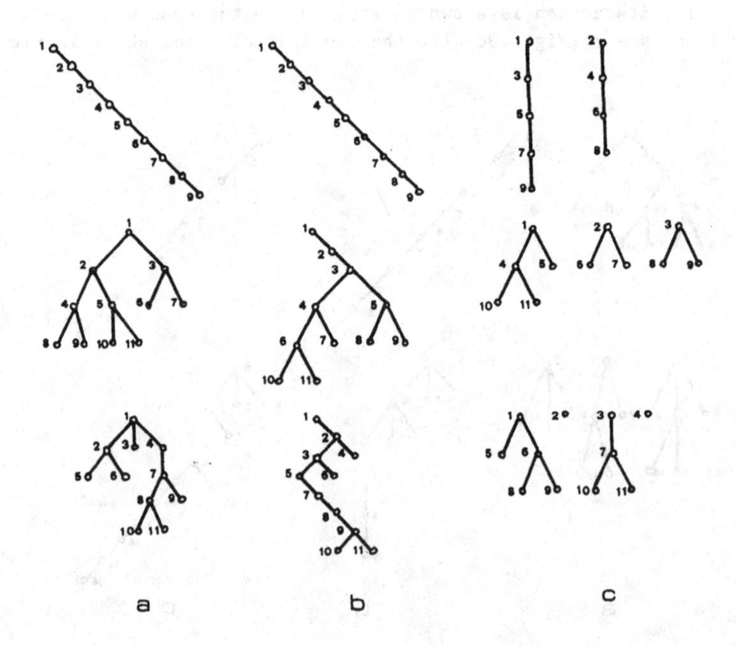


Fig. 1.10

We are not going to discuss similar operations obtained by adding (or subtracting) an arbitrary integer number c instead of one unit to function S, because the resulting operation is a repetition of operations 1.8 and 1.9.

Consider now the parabolic growth operator

$$(1.10) \qquad \psi(n) = \sqrt{\varphi(n)}$$

This operation "compresses" the number of levels of the hierarchy and, therefore, the number of subordinated vertices in each level increases "quadratically", so to speak.

In Fig.1.11, we show the action of this operator on the same initial graphs as those used in Fig.1.8.

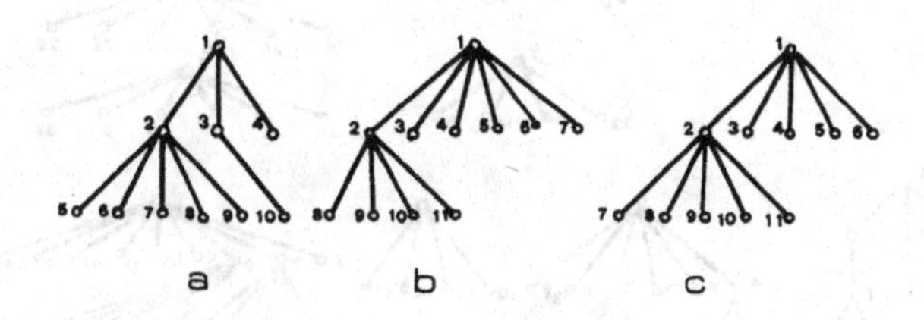

Fig. 1.11

Also arithmetical operations on the arguments of a function S transform it into a new function S, these too may therefore be seen as unary operations. With them, the definition field of function S, i.e. the set of numbers 1,2,...N, may change in such a manner,that the new definition field is not necessarily a set of integer numbers 1,2,...K. It is then necessary to complete definitions of these operations in a convenient way.

The evolution or reorganization operation into smaller units is defined as

$$(1.11) \qquad \psi(n) = \varphi\left(\frac{n}{K}\right), \qquad K > 1$$

under condition

$$\varphi\left(\frac{n}{K}\right) = 1 \qquad \text{for } \frac{n}{K} < 1$$

due to the fact that functions S are not generally defined in zero. In Fig.1.12, we show the action of operation 1.11 on the same three trees (Fig.1.2a) which were used in Fig.1.8. In Fig.1.12b, we consider the case when K = 2, and in Fig.1.12c, K = 3.

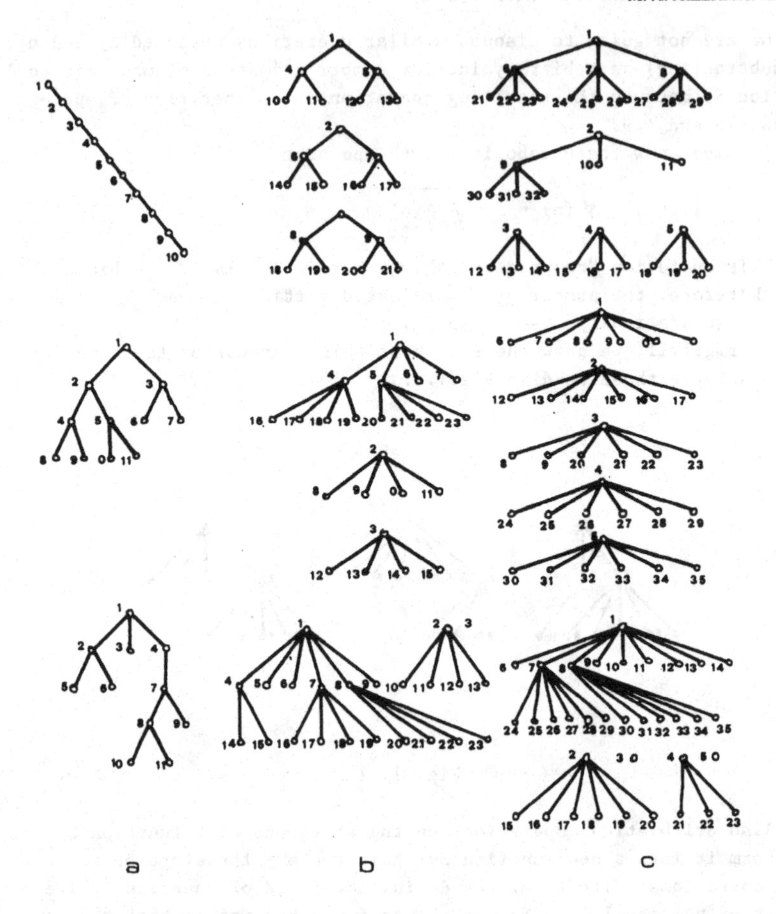


Fig. 1.12

A comparison between Figs.1.12 and 1.8 shows that operation 1.11 acts approximately in the same way as Fig.1.6, that is, it generates several trees, which take their roots from the vertices of the highest levels of the original tree and distributes the other subordinated vertices among these.

In addition, if no special condition prevents it, operation 1.11 increases the total number of vertices by a factor K with respect to the initial tree. Therefore, as regards an administrative structure, this operation describes a division in the structure accompanied by a growth in the number of staff.

The contraction or reorganization operation into larger units is defined as

$$(1.12) \qquad \psi(n) = \varphi(Kn), \qquad K = 2,3,\ldots$$

The operation reduces the number of the tree's vertices and, at the same time, it produces a stretching on the tree. The action of the operation is shown in Fig.1.3. Operation 1.12 may break condition 1°; in this case, the value of the result of the operation is defined on the basis of the fundamental definition.

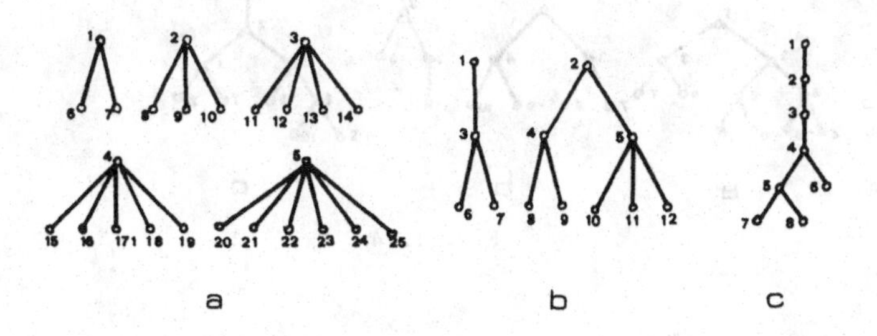


Fig. 1.13

The two following operations define a new subordination system for simple vertices, with possible jumps and transitions to higher ranks

$$(1.13) \qquad \psi_1(n) = \varphi(n \quad c), \qquad c = 1,2,3,\ldots,$$

$$(1.14) \qquad \psi_2(n) = \varphi(n + c) \qquad c = 1,2,3,\ldots,$$

Operation 1.13, with $n = 1,2,\ldots$, needs a further definition, as (0) is not defined. In this case, we will assume that $\varphi(0) = \varphi(1) = 0$. Further, this operation increases the total number of vertices from N to N + c. As it will sometimes be necessary to use the same operation, keeping, however, the number of vertices, we will consider also a possible variant of the definition (1.13). It is given by

$$\varphi(N+1) = \varphi(N+2) = \ldots\ldots = \varphi(N+c) = 0$$

The meaning of (1.13) is that each subordinated vertex is now subject to a new leading vertex, that to which vertices with a c-times smaller index were previously subordinated. It is then natural that, if c is small and the tree has a relatively large number of branches, a great part of the vertices will keep the same kind of subordination as before, and only "extreme" vertices of the branch will be reorganized.

Operation 1.14 acts in a similar way to (1.13) but with a downwards, instead of an upwards movement in the levels of the hierarchy. Two examples of the action of operations 1.13 and 1.14 are shown in Fig.1.14b and 1.14c.

When $N > c$, operation 1.14 reduces the total number of vertices from N to N - c.

Let us consider now a function p(n) which takes values only from the set $\{-1,0,+1\}$, and introduce operation

$$(1.15) \qquad \psi(n) = \varphi(n) + p(n)$$

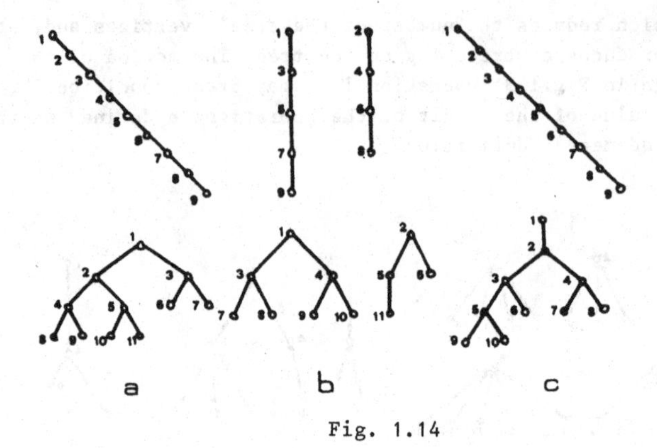


Fig. 1.14

Comparing this expression with (1.8) and (1.9), it can immediately be seen that operation 1.15 produces local rearrangements of the subordination in tree $\varphi$ (n); position and direction of these rearrangements are wholly determined by the concrete form of function p(n). We can therefore refer to p(n) as the control function of the rearrangement. Examples of the action of this operation are shown in Fig. 1.15.

In Fig. 1.15a we fix graph $\varphi$ (n); in Fig. 1.15b, we define p(n) in various manners; and in Fig. 1.15c, we report the resulting graphs.

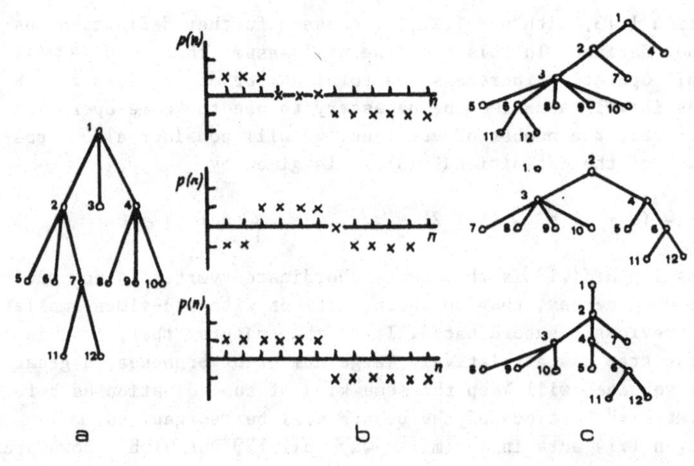


Fig. 1.15

The cutting operation is defined as

$$(1.16) \qquad \psi(n) = p(n) \cdot \varphi(n)$$

where, in this case, $p(n) = 1/n$.

It is obvious that with this operation each forest (or, in particular, each tree) is divided into a set of non-connected vertices.

Finally, we should say here that, even with the limited set of the above-mentioned operations, it is always possible to form new operations as functions of functions. By combining the operations in this way, it becomes possible to form trees with specific assigned properties.

For example, the binary tree can be obtained with the aid of the following four operations:

1. $\varphi_1(n) = 0$; function $\varphi_1(n)$ specifies a diffuse graph
2. $\varphi_2(n) = \varphi_1(n) + 1$; the diffuse graph is transformed into a fan graph (Fig.1.7a)
3. $\varphi_3(n) = n\,\varphi_2(n)$; the fan is transformed into a chain graph (Fig.1.4)
4. $\varphi_4(n) = \varphi_3(n)/2$ is the binary tree (Fig.1.9a)

Binary or more complicated operations

Let us begin now with the descriptions of binary operations, where two functions S: $\varphi(n)$ and $\pi(n)$ take part.

The additional definition 1.5, previously introduced, can be extended to binary operations, as well as to n-ary operations or still more complex operations, which will be introduced later. Thus, it follows that as a result of a binary operation, we still have a function S.

The arithmetical mean operation is defined as follows

$$(1.17) \qquad \psi(n) = \varphi(n) + \pi(n) \quad /2$$

where $\lfloor a \rfloor$ denotes the integer part of a.

In Fig. 1.16, we give an example in which $\varphi$ has a chain structure (Fig. 1.16a), whereas $\pi$ has a fan structure (Fig. 1.16b).

The operation generates a binary tree (Fig. 1.16c). The "mean" character of the operation is even more evident in the example in Fig. 1.17.

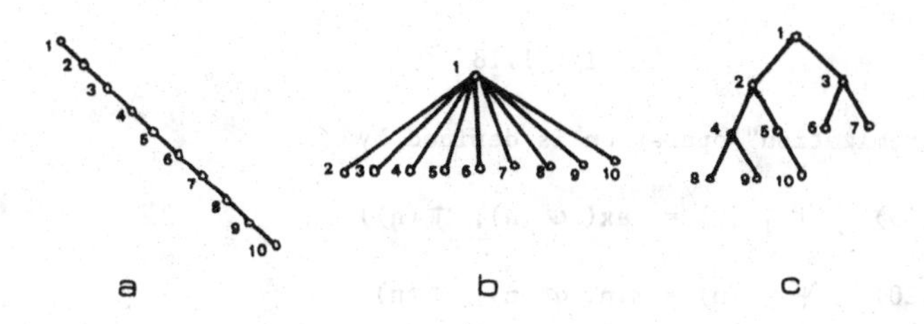


Fig. 1.16

where $\varphi$ and $\pi$ have been formed in the same manner but have a "different inclination" (Fig.1.17a and b); the result is a tree without inclination (Fig.1.17c)

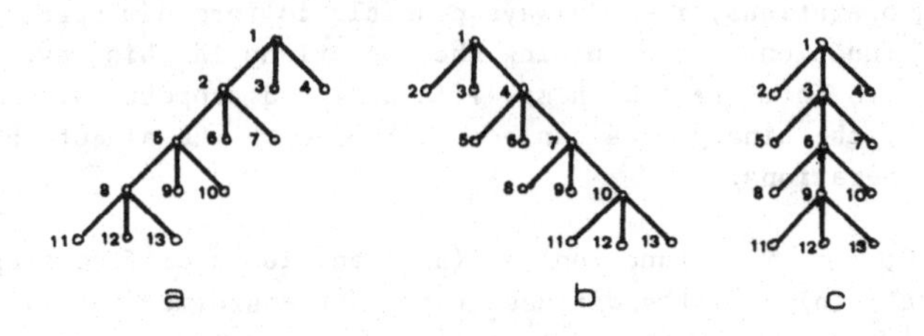


Fig. 1.17

The goemetric mean operation is defined as follows

$$(1.18) \qquad \Psi(n) = \quad \varphi(n)\ \pi(n)$$

In Fig.1.18, we show the result of this operation applied to the same two graphs used in the example in Fig.1.16

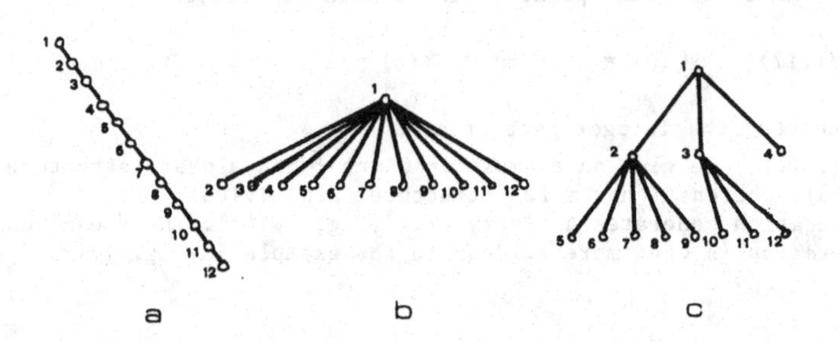


Fig. 1.18

The "extremization" operation is defined by

$$(1.19) \qquad \Psi_1(n) = \max(\varphi(n), \pi(n))$$

or

$$(1.20) \qquad \Psi_2(n) = \min(\varphi(n), \pi(n))$$

These operations applied to the two trees form just one tree with the following general trend: if the first extends "in height", i.e. it has many levels, whereas the other extends "in width", i.e. it has many elements on each level, then, 1.19 forms a tree "extended in height", while 1.20 forms a tree "extended in width".

We can also introduce operations acting on r functions S.

An example of such an operation is the arithmetical mean on r function S

$$(1.21) \qquad \Psi(n) = \left[ \sum_{i=1}^{r} \varphi_i(n)/r \right]$$

Function $\Psi(n)$ is defined with the integers 1,....,N, where N is equal to the minimal number of nodes among graphs $\varphi_i(n)$. The result of operation 1.21 is shown is Fig.1.19, where $\Psi(n)$ (Fig.1.19c) is formed starting from the four graphs $\varphi_i(n)$, (i = 1,2,3,4) shown in Fig.1.19 a,b,c,d.

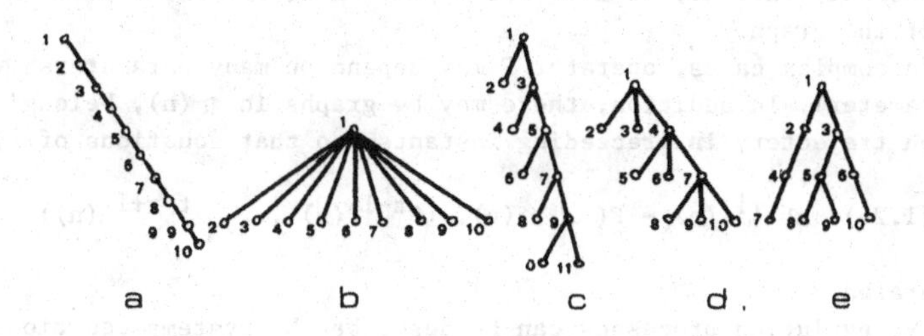


Fig. 1.19

If we consider the case in which a non-connected set of nodes must be transformed into a structure where the number of subordinated vertices tends to increase when we pass from the highest to the lowest levels of the hierarchy, such operation can be performed by collecting first the scattered points in a binary tree, and then transforming this into the required one. This collection of operations may be written as follows:

$$(1.22) \qquad \Psi(n) = \sqrt{\xi(n)}$$

where

$$\xi(n) = n \; \frac{\varphi(n) + 1}{2}$$

It is also useful to introduce the weighted mean operation, defined by

$$(1.23) \qquad (n) = \frac{K_1 \varphi_1(n) + \ldots\ldots. + K_r \varphi_r(n)}{K_1 + \ldots\ldots. + K_r} ,$$

where $K_i$ are positive numbers.

## 1.3 Graph Dynamics Equations for Functions S. First Order Equations.

Equations of graph dynamics defining graph trajectories may be seen as iterative procedures written in terms of functions and operations. Thus, for instance, graph dynamics equations can be written in the following form

$$(1.24) \quad \varphi^{t+1}(n) = F(\varphi^{t}(n)),$$

where F is a unary operation (containing or not a variable parameter p(n) ), or in the form

$$(1.25) \quad \varphi^{t+1}(n) = F(\varphi^{t}(n), \pi(n)) \quad ,$$

where F is a certain binary operation on $\varphi$ and $\pi$ ; the function S $\pi$ (n) which is constant, that is, it does not depend on t, has the role of a parameter of the graph.

In more complex cases, operator F may depend on many parameters $\pi$(n) (vector-parameter). In addition, there may be graphs in $\pi$(n), belonging to the graph trajectory in preceeding instants, so that equations of type

$$(1.26) \quad \varphi^{t+1}(n) = F(\varphi^{t}(n), \varphi^{t-1}(n) \ldots . \varphi^{t-r+1}(n))$$

may be generated.

Finally, evolution processes can be described by systems equations of the form

$$(1.27) \quad \varphi^{t+1}(n) = F(\varphi^{t}(n), \psi^{t}(n)), \psi^{t+1}(n) = = F(\varphi^{t}(n), \psi^{t}(n))$$

or more complex systems, if operators $F\varphi$ and $F\psi$ depend on constant parameters or on graphs appearing in preceeding instants.

Obviously, as well as operators, "initial condition" must also be given, i.e. for equations 1.24 or 1.25 on the initial graph $\varphi^{\circ}(n)$, and for equations 1.26 or system 1.27, a group of initial graphs.

Unary operator

Let us consider equation 1.24 with the assumption that F is a unary operation, not containing variable parameter p(n). According to the introduced definition, a graph $\varphi^{*}(n)$ satisfying condition

$$(1.28) \quad \varphi^{*}(n) = F(\varphi^{*}(n))$$

is called equilibrium graph.

The solution of the functional equation 1.28 establishes the equilibrium graphs of the problem 1.24.

Theorem 1.1

If F is an increasing monotonic function, then the solution of equation 1.28 always exists and is determined univocally once the initial graph $\varphi^{\circ}(n)$ has been assigned.

Proof

For each value $n_0$, three cases are possible:

1. $F(\varphi^0(n_0)) = \varphi^0(n_0)$ : obviously in this case $\varphi^0(n_0) = \varphi^*(n_0)$

2. $F(\varphi^0(n_0)) < \varphi^0(n_0)$ : let $\varphi^0(n_0) = K_0$, $F(\varphi^0(n_0)) = K_1$, $F(F(\varphi^0(n_0))) = K_2$, etc.; it is sufficient to note that the sequence $K_0, K_1, K_2 \ldots$ decreases monotonically and that for each i, $K_i \geqslant 0$.

3. $F(\varphi^0(n_0)) > \varphi^0(n_0)$

Similarly, it can be verified that the sequence $K_0, K_1, K_2, \ldots$ is monotonic and non-decreasing and that, on the other hand, has a superior limit of the number $n_0 - 1$.

It is obvious that, if for some i, $K_i = K_{i+1}$ , then, for every $s > i, K_s = K_i$.

As $0 \leqslant \varphi(n) < n-1$ and $\varphi(n)$ is an integer number, and since the sequence $K_0, K_1, K_2, \ldots$ is monotonic, it is clear that $\varphi^0(n)$ being assigned, only a single function S, $\varphi^*(n)$ exists, such that $\varphi^*(n) = F(\varphi^*(n))$.

Theorem 1.2

If F is a one-directional operation, then the solution of equation 1.28 always exists and is determined univocally, $\varphi^0(n)$ being assigned. Yet, if F is a strictly one-directional operation, then only the two following functions are possible:

(a) $\varphi^*(n) = 0, \quad (n = 1,2,\ldots\ldots,N)$ ,

(b) $\varphi^*(n) = n - 1 (n = 1,2,\ldots\ldots,N)$

Proof

If F is one-directional, then for each n $F(\varphi(n)) \geqslant \varphi(n)$ or $F(\varphi(n)) \leqslant \varphi(n)$. For instance, let $F(\varphi(n)) \geqslant \varphi(n)$ for each n, then for any $n_0$ the sequence

$$\varphi^0(n_0),\ F(\varphi^0(n_0)),\ F(F(\varphi^0(n_0))),\ldots\ldots$$

is monotonic non-decreasing, and it has a superior limit of $n^0 - 1$. Consequently, there is $\varphi^*(n)$, which is the solution of 1.28.

If F is strictly one-directional, then for any $n_0$ we have a monotonic increasing sequence limited by the number $n_0 - 1$, which is in fact its limit. Case $F(\varphi(n)) \leqslant \varphi(n)$ may be treated in the same manner.

In relation to the fact that the number of different graphs with N nodes is finite, the existence of the solution to equation 1.28 is equivalent to the absence of cycles in the correspondent graph trajectory and, therefore, the conditions for the existence of a solution - given in theorems 1 and 2 - are equivalent to the condition of absence of cycles.

All unary operations introduced in Part 1 are one-directional; thus, considering theorems 1 and 2, the solution of equation 1.28 always exists for them.

We will state that an equilibrium-approaching process is globally convergent to equilibrium if $\varphi^*(n)$ does not depend on the initial graph $\varphi^0(n)$. It is not diffecult to see that the condition of one-directionality is also the condition of global convergence to equilibrium.

We now introduce the notion of distance between graphs.

Take $\rho(\varphi,\Psi)$ as distance between two graphs given by functions S $\varphi(n)$ and $\Psi(n)$ the quantity

$$(1.29) \quad \rho(\varphi,\Psi) = \sum_{n=1}^{N} (\varphi(n) - \Psi(n))^2 / N^2$$

Theorem 1.3

If in equation 1.24 F is a one-directional operation, then a process described by this equation converges monotonically to equilibrium with respect to $\rho$.

Proof

If F is a one-directional operation, then $\varphi^0(n) \leqslant \varphi^*(n)$, o$\varphi^0(n) \geqslant \varphi^*(n)$, where $\varphi^*(n)$ is the solution of equation 1.28 for this process.

From F's one-directionality, it follows that, if $\varphi^0(n) \leqslant \varphi^*(n)$, then $\varphi^0(n) \leqslant \varphi^t(n) \leqslant \varphi^*(n)$ and $\varphi^i(n) \leqslant \varphi^j(n) \leqslant \varphi^*(n)$ for $i \leqslant j$. Hence, it follows directly that

$$\rho(\varphi^i,\varphi^*) > \rho(\varphi^{i+1},\varphi^*)$$

Similarly, for $\varphi^0(n) \leqslant \varphi^*(n)$ we have

$$\rho(\varphi^i,\varphi^*) < \rho(\varphi^{i+1},\varphi^*)$$

The monotonicity with respect to $\rho$ means that any successive graph is, in a certain sense, nearer to the preceeding graph than to the final one. In Fig.1.20, three examples of trajectories formed by recursive relations of the form(1.24)

$$\text{(Fig.1.20a)} \quad \varphi^{t+1}(n) = \varphi^t(n) \dot{-} 1$$

$$\text{(Fig.1.20b)} \quad \varphi^{t+1}(n) = \varphi^t(n) + 1$$

$$\text{(Fig.1.20c)} \quad \varphi^{t+1}(n) = \left[\sqrt{\varphi^t(n)}\,\right]$$

In all these three cases, we have used the same initial graph $\varphi^0(n)$.

In case(1.20a),this graph, after five steps, has been transformed into a set of isolated nodes, that is, it is completely restructured.

**In case b, after six steps, the initial graph has been reduced to a chain, whereas in case c, after two steps, it has been reduced to a fan.**

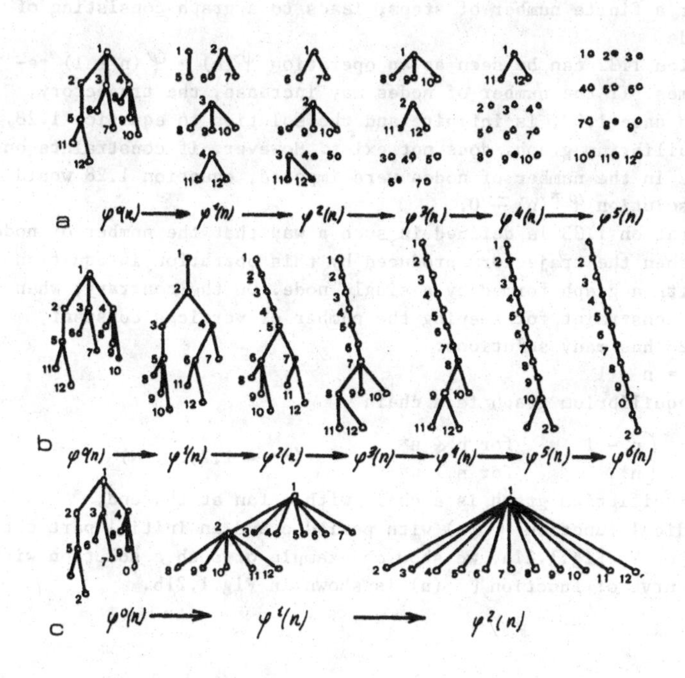


Fig. 1.20

Operations with arguments

We have already introduced the following operations with the arguments of a function S

$$(1.30)\quad \Psi(n) = \varphi\left(\left[\frac{n}{k}\right]\right) \qquad (k = 1,2,3,\ldots.)$$

$$(1.31)\quad \Psi(n) = \varphi(kn) \qquad (k = 1,2,3,\ldots.)$$

$$(1.32)\quad \Psi(n) = \varphi(n \dot{-} c) \qquad (k = 1,2,3,\ldots.)$$

$$(1.33)\quad \Psi(n) = \varphi(n + c) \qquad (k = 1,2,3,\ldots.)$$

and have described the properties of function $\Psi(n)$ thus obtained. All four operators are neither monotonic nor one-directional: consequently they do not satisfy the conditions of theorems 1.1 and 1.2.

From a certain initial graph $\varphi^0(n)$, operation 1.30 produces an infinite trajectory of graphs having an unlimited growth in the number of nodes. When constraints are imposed, which do not permit a growth in the number of nodes, after a finite number of steps, the trajectory arrives at a "scattered"graph $\varphi^*(n) \equiv 0$, which is actually the equilibrium graph according to 1.28.

Whatever the initial graph, operation 1.31 produces a trajectory which, after a finite number of steps, leads to a graph consisting of a single node.

Operation 1.32 can be seen as an operation $\Psi(n) = \varphi(n - 1)$ repeated c-times. If the number of nodes may increase, the trajectory, similarly to case 1.30, is infinite and the solution to equation 1.28, i.e. the equilibrium graph, does not exist. However, if constraints on the increase in the number of nodes were imposed, equation 1.28 would allow only solution $\varphi^*(n) \equiv 0$.

If operation 1.33 is defined in such a way that the number of nodes decreases, then the trajectory produced by this operation in equation 1.28 ends with a graph formed by a single node. On the contrary, when we impose a constraint for keeping the number of vertices constant, equation 1.28 has many solutions:

1) $\varphi^*(n) = n - 1$

i.e. the equilibrium graph is a chain

2) $\varphi^*(n) = \begin{cases} n - 1 & \text{for } n \leqslant n^* \\ n^* & \text{for } n > n^* \end{cases}$ $(n^* = 1, 2, , \ldots N)$

i.e. the equilibrium graph is a chain with a fan at the end;

3) Any periodical function $\varphi^*(n)$ with period c and an initial part not longer than c. In Fig.1.21a, we show an example of such a solution with c = 3. The curve of function $\varphi^*(n)$ is shown in Fig.1.21b.

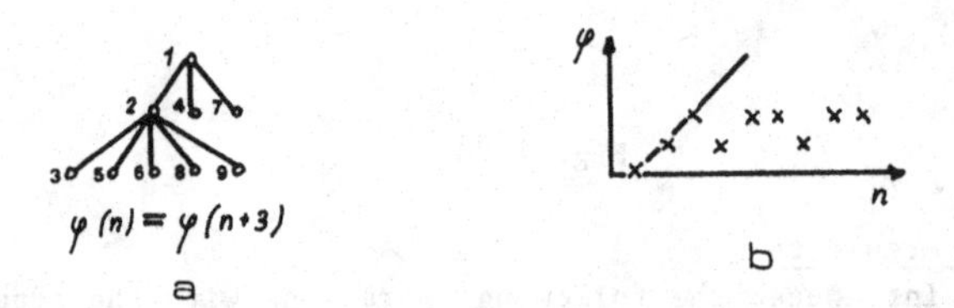


Fig. 1.21

Binary Operator

Let us consider now equation 1.25 and assume that F is one of the two binary operators previously introduced

$$(1.34) \qquad F_1 = \left[(\varphi(n) + \pi(n))/2\right]$$

$$(1.35) \qquad F_2 = \left[\sqrt{\varphi(n)\,\pi(n)}\right]$$

We will also assume that the number of nodes in graph $\varphi^0(n)$ is not smaller than those in graph $\pi(n)$.

As before, an equilibrium graph is determined by relation

$$(1.36) \quad \varphi^*(n) = F(\varphi^*(n), \pi(n))$$

where $F = F_1$ or $F = F_2$, whereas $\pi(n)$ is a specified graph.

We will consider now two sets of graphs $S^*_{\pi(n)}$ and $S^{**}_{\pi(n)}$, given as follows

$$(1.37) \quad \begin{cases} S^*_{\pi} = \{\varphi(n) \;/\; \pi(n) - \varphi(n) = 0 \text{ or } 1\} \\ S^{**}_{\pi} = \{\varphi(n) \;/\; \pi(n) - \varphi(n) = 0 \text{ or } 1 \text{ or } 2, \text{ or } \pi(n)\} \end{cases}$$

Set $S^*_{\pi}$ contains graph $\pi(n)$ as well as all graphs given by functions which for any n are equal or less than one unit with respect to $\pi(n)$, whereas $S^{**}_{\pi}$ contains all graphs resulting from functions which for any n coincide with $\pi(n)$ or are less than 1 or 2 units with respect to $\pi(n)$.

Theorem 1.4

The solution of equation 1.36, for $F = F_1$ or $F = F_2$, always exists and belongs to $S^*_{\pi}$ for $F = F_1$ and to $S^{**}_{\pi}$ for $F = F_2$. This solution is univocally determined by initial graph $\varphi^0(n)$.

Proof

If $F = F_1$ and $\varphi^0(n) \geqslant \pi(n)$, the solution of 1.28 is then $\varphi^*(n) = \pi(n)$. If for a given $n_0$, $\varphi^0(n_0) < \pi(n_0)$, it is easily seen that $\varphi^*(n_0) = \pi(n_0) - 1$, then, considering the adopted quantization method, we take the nearest integer. However, if $F = F_2$, it is then clear that for the n's for which $\varphi^0(n) = 0$, $\varphi^*(n) = 0$, whereas in points where $\varphi^0(n) \geqslant \pi(n)$, we have $\varphi^*(n) = \pi(n)$. We must now consider the points in which $0 < \varphi^0(n) < \pi(n)$. If $\varphi^i(n) = \pi(n) - 1$, then

$$\varphi^{i+1}(n) = \left[\sqrt{(\pi(n) - 1)\,\pi(n)}\right] = \pi(n) - 1$$

If $\varphi_i(n) < \pi(n) - 1$, then we obtain $\varphi^{i+1}(n) \leqslant \pi(n) - 2$ because $\sqrt{K^2 - 2K} < K - 1$.

Classes $S^*_{\pi}$ and $S^{**}_{\pi}$ consist of all the functions taking the above-mentioned values.

All the graphs belonging to set $S^*_{\pi}$ differ little from graph $\pi(n)$ and, except for this little difference, we can state that for equation 1.25 the equilibrium graph is given by graph $\pi(n)$ when $F = F_1$. When $F = F_2$, the situation is more complicated. If $\varphi^0(n)$ is a tree (and not a forest) then $\varphi(n) = 0$ only at the root (for $n = 1$) and for all the n's case $\pi(n) - \varphi(n) = \pi(n)$ is impossible. Also here the equilibrium graph coincides with $\pi(n)$ or it is very near to it. The situation is different when $\varphi^0(n)$ is a forest. In this case the equilibrium graph may differ substantially from $\pi(n)$: $\pi(n)$ may be a forest, even if $\pi(n)$ is a tree; function $\varphi(n)$ differs from $\pi(n)$ precisely in the points representing the roots of the forest's trees. We can now characterize the properties of graph trajectories.

Theorem 1.5
For any initial graph $\varphi^0(n)$, the graph trajectory tends monotonically, with respect to $\rho$, to an equilibrium graph belonging to $S^*_\pi$, if $F = F_1$, or to $S^{**}_\pi$, if $F = F_2$. If $F = F_1$, the number of steps necessary to establish the equilibrium does not exceed $[\ln N / \ln 2] + 1$; on the contrary, if $F = F_2$, then the number of steps does not exceed $[\ln(\ln N / \ln 2)/\ln 2] + 1$, where N is the number of nodes in graph $\varphi^0(n)$.

Proof
The monotonicity with respect to $\rho$ results from the inequality $|\varphi^i(n) - \varphi^*(n)| \leq |\varphi^{i+1}(n) - \varphi^*(n)|$. This inequality is obvious for sequences $K_{i+1} = K_i + c/2$ and $K_{i+1} = \sqrt{K_i c}$ with integer c. The evaluation of the number of steps (in case $F = F_1$) follows from the inequality $N/2^K < 1$ being $\max(\varphi(n) = 1 - \pi(n)) \leq N$. In case $F = F_2$, if $\varphi^0(n) = 1$ or $\pi(n) = 1$, the evaluation derives from the inequality $N^{2^{-K}} < 2$.

It is easy to show that, if $\varphi^0(n) > 1$ and $\pi(n) > 1$, less than k steps are necessary. From theorem 1.5 it follows that for binary operation each initial graph is translated, after a finite number of steps, monotonically with respect to $\rho$ into any other assigned graph $\pi(n)$ (or one very close to it). In this sense, equation 1.25, for $F = F_1$ describes a possible monotonic variation procedure of a hierarchical structure, from any initial structure to any other assigned.

In Fig. 1.22, we show an example of trajectory, corresponding to equation 1.25, when the initial graph is a chain graph (Fig. 1.22a) or a fan graph (Fig. 1.22b), and the graph acting as parameter $\pi(n)$ is a binary tree.

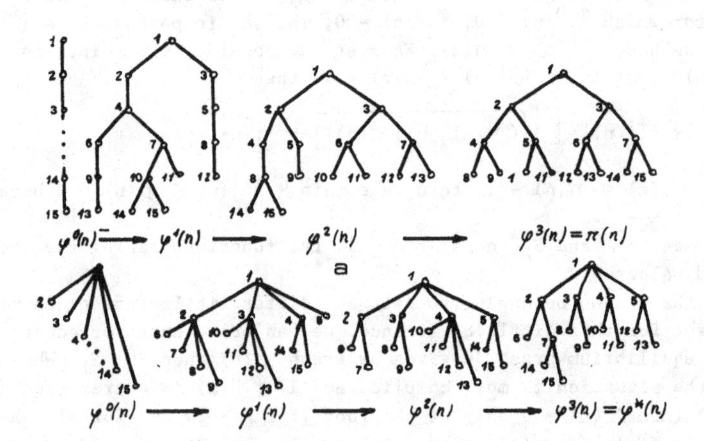


Fig. 1.22

In Fig. 1.23, we show the trajectories for the same initial graphs and for the same parameters, but for $F = F_2$.

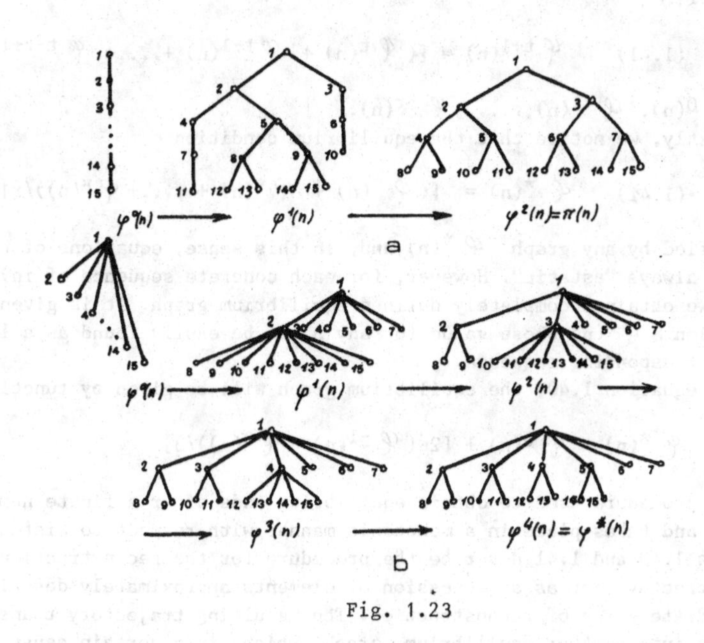


Fig. 1.23

In conclusion, we observe the following. If we introduce another quantization method and define the binary (non-commutative) operation as

$$(1.38) \qquad \varphi \boxplus \psi = \begin{cases} [\varphi(n) + \psi(n))/2] & \text{for } \varphi(n) > \psi(n) \\ [\varphi(n) + \psi(n))/2] + 1 & \text{for } \varphi(n) < \psi(n) \end{cases}$$

or as

$$(1.39) \qquad \varphi \square \psi = \begin{cases} [\sqrt{\varphi(n)\ \psi(n)}] & \text{for } \varphi(n) \geqslant \psi(n) \\ [\sqrt{\varphi(n)\ \psi(n)}] + 1 & \text{for } \varphi(n) < \psi(n) \end{cases}$$

then, equation 1.25 translates any initial graph into $\psi(n)$ (due to operation 1.38) and any initial tree into $\psi(n)$ (due to operation 1.39).

<u>Higher order equations and systems of equations</u>

Equations of a higher order than the first one require the presence of a binary or r-ry operation.

We will limit ourselves to considering an example of a second order equation

$$(1.40) \qquad \varphi^{t+1}(n) = [(\varphi^{t}(n) + \varphi^{t-1}(n))/2]$$

having been assigned $\varphi^{0}(n)$, and $\varphi^{-1}(n)$, and of an analogous equation of order r

$$(1.41) \qquad \varphi^{t+1}(n) = [(\varphi^{t}(n) + \varphi^{t-1}(n) + \ldots + \varphi^{t-r+1}(n))/r$$

given $\varphi^{0}(n), \varphi^{-1}(n), \ldots, \varphi^{-r}(n)$.

Firstly, we notice that the equilibrium condition

$$(1.42) \qquad \varphi^{*}(n) = [(\varphi^{*}(n) + \varphi^{*}(n) + \ldots + \varphi^{*}(n))/r]$$

is satisfied by any graph $\varphi^{*}(n)$ and, in this sense, equations of these kind are always "astatic". However, for each concrete sequence of initial graphs, we obtain a completely defined equilibrium graph. It is given by function S $\varphi^{*}(n)$ whose value for any n can be easily found as a limit of the corresponding sequence.

For equation 1.40, the equilibrium graph will be given by function

$$\varphi^{*}(n) = \varphi^{0}(n) + [2(\varphi^{-1}(n) - \varphi^{0}(n))/3]$$

The procedure used to obtain equilibrium ends after a finite number of steps and takes place in a monotonic manner with respect to distance $\rho$. Equations 1.40 and 1.41 describe the procedure for the reconstruction of the structure, or as a succession of elements approximately describing the immediate stage of reconstruction. The resulting trajectory translates the graph into another equilibrium graph, which, in a certain sense, will take into consideration the structures specified in the initial sequence $\varphi^{-k}(n)$ $(k = 0, 1, \ldots, r)$.

So far, we have discussed only examples with simple operators. To write more complex equations, it is necessary to formalize some general principles or "life rules" of the structure, which determine the relations in each step of its reconstruction. The solution of the equation means the possibility of foreseeing the results of equations of these general laws. The aim of this chapter is to propose an apparatus for investigations of this kind and not for the study of internal laws, characteristic of structures described by this apparatus. For this reason, we are not going to give further information about the suitability of the language formulated for a single class of integer functions or about examples confirming this.

This circumstance seems even clearer when we analyze systems of graph dynamics equations.

We limit ourselves to considering the description and the analysis of a single significant problem which leads to a system of graph-dynamics equations. We assume that there are two independent structures, which are respectively described by functions $\varphi(n)$ and $\psi(n)$.

In addition, we assume that there are some constraints on structure $\varphi(n)$ so that it approaches structure $\psi(n)$, keeping, however, its initial properties. Such constraints may be, for example, of the following type:

$$(1.43) \qquad \varphi^{t+1} = [k\, \varphi^{t}(n) + \psi(n) \;/\; (k + 1)] \quad (k = 1, 2, 3, \dots)$$

If the structure $\psi(n)$ remains constant, 1.43 describes the iterative structure of the transition from $\varphi(n)$ to $\psi(n)$; the larger the number k, the smaller is the change of $\varphi(n)$ at any iterative step.

However, we assume that also structure $\psi(n)$ approaches $\varphi(n)$, with the same constraints and laws

$$(1.44) \qquad \psi^{t+1}(n) = [k\,\psi^{t}(n) + \varphi(n)) \;/\; (k + 1]$$

Then, such a "mutual approach" of the structures is described by the system of equations

$$\varphi^{t+1}(n) = [k\, \varphi^{t}(n) + \psi^{t}(n)) \;/\; (k + 1)]$$

$$\psi^{t+1}(n) = [k\, \psi^{t}(n) + \varphi^{t}(n)) \;/\; (k + 1)]$$

with "initial structures" $\varphi^{0}(n)$ and $\psi^{0}(n)$.

Coefficient k, in these equations, has the role of indicating the tendency of each structure, during reconstruction, to keep its "own" structure.

The equilibrium condition, in this problem, is determined by the system of functional equations

$$(1.45) \qquad \varphi^{*}(n) = [(k\, \varphi^{*}(n) + \psi^{*}(n)) \;/\; (k + 1)],$$

$$\psi^{*}(n) = [(k\, \psi^{*}(n) + \varphi^{*}(n)) \;/\; (k + 1)]$$

It is not difficult to see that any pair of coincident functions $\varphi^{*}(n) = \psi^{*}(n)$ satisfies system 1.45. This means that, as a result of the action of system 1.44, the two initial graphs will become equal to each other, whereas only the character of the trajectories will depend on k.

As an example of trajectories describable by system 1.44, we consider the two cases shown in Figs. 1.24 and 1.25.

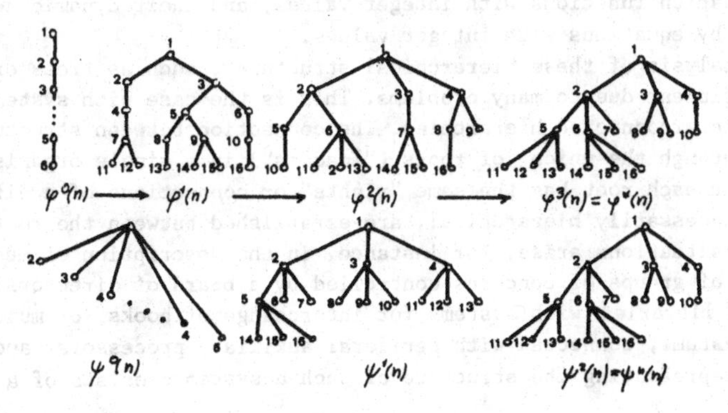

Fig. 1.24

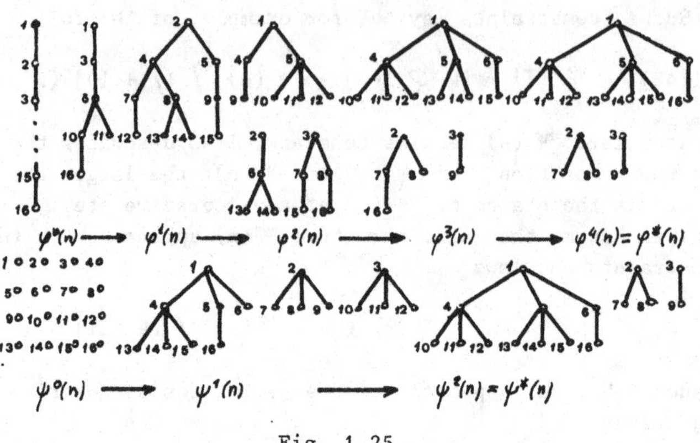


Fig. 1.25

In both cases, we impose K = 2; as initial graph $\varphi^0(n)$ we took a chain, while as initial graph $\psi^0(n)$ we took a fan (Fig. 1.24) in one case, and a set of non-connected nodes (Fig. 1.25) in the other.

In relation to the examples studied, we observe in conclusion, that when we describe a structure with a function , new results have nothing to do with the use of the mathematical apparatus of integer equation theory, but with the analysis of the equations' content and of their solution.

Thanks to this, the "life" of a hierarchical structure becomes the subject of formal analysis and makes it possible to anticipate the results in the change of a structure, when the changes themselves are locally specified.

## 2. SUBSTITUTION OPERATOR AND COMPLEX HIERARCHICAL GRAPHS

In the first chapter we considered variation processes in the life of hierarchical structures, formally represented by directionally numbered graphs: trees or forests. The static structure of such systems may be described by subordination functions with integer values, and their dynamic development with time, by equations with integer values.

The analysis of these hierarchical structures, such as trees or forests, is not sufficient due to many problems. This is the case with systems containing many interconnected hierarchies. The connection between structures can be shown through the union of roots ("leaders") in a single organism, in which either each root has the same "rights", or connections of a different kind, not necessarily hierarchical, are established between the roots.

These situations arise, for instance, in the description of menagment structures of groups of concerns controlled by a board of directors, or networks of libraries with systems for interchange of books, or multiprocessor computer systems, connected with periferal auxiliary processors, and so on. The graph representing the structure of such a system consists of a set of separate trees with additional connectors among their roots.

In this chapter, we will introduce graphs of this kind which are called complex hierarchical graphs, in short, CH-graphs.

In chapter I, the subordination function $x_m = \varphi(x_n)$ was introduced, which produces a univocal correspondence of a finite set of integers [O, N] in itself and satisfies condition

$$(2.1) \qquad 0 \leq \varphi(x_n) < x_n, \qquad x_n \in [0, N]$$

We will show that the elimination of constraint 2.1 and the assumption that $\varphi(x_n)$ is a function which makes a numberable set of elements correspond in itself, allows the description of a CH-graph class. In paragraph 2.1, we will introduce the notion of function S, defined using an arbitrary numberable set X of elements, and we will prove the existence of a biunivocal correspondence between the set of functions S and the set of CH-graphs. In paragraph 2.2, we will present a class of substitution operators mantaining "the complexity" of the initial graph and we will investigate its fundamental properties.

## 2.1 Structure of CH-graphs generated by functions S

Let us consider now an arbitrary set X of different elements $x \in X$ and an arbitrary function $\varphi(x)$ which makes the set X correspond to itself: $\varphi : X \longrightarrow X$. In correspondence to function $\varphi(x)$, we define an oriented graph $\Gamma_\varphi$, as follows: for $x, y \in X$ the expression $y = \varphi(x)$ means that the graph contains the adjacent vertices classified by x and y and that the edge xy is directed from x to y. As the X set is finite and the correspondence is into itself, each connected component of graph $\Gamma_\varphi$ includes a cycle (in particular a cycle of length 1, i. e. an encircled vertex). Each connected component of graph $\Gamma_\varphi$ contains more than one cycle. This means that to each vertex of a cycle only a graph without cycles, that is a tree, may be attached.

Thus, graph $\Gamma_\varphi$ may be an arbitrary collection of connected components of the four following types (Fig. 2.1):

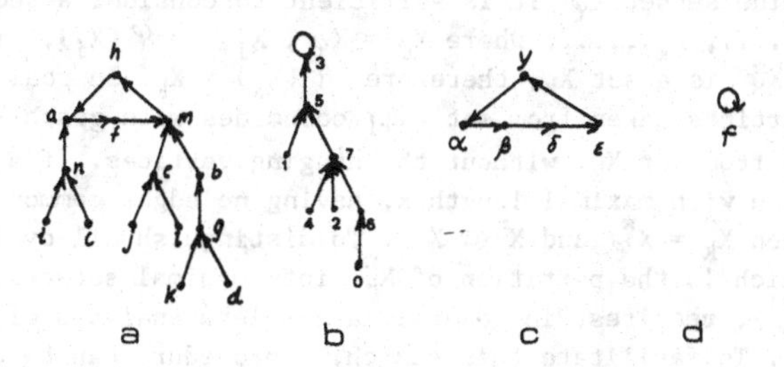


Fig. 2.1

oriented cycles with trees whose roots are vertices of the cycle (Fig. 2.1a) and its three particular cases - a tree with its root encircled (Fig. 2.1b), an isolated cycle (Fig. 2.1c) and an isolated vertex (Fig. 2.1d).

When the set X is denumerable and when the subset $X'_\varphi \subsetneq X$ of elements $x \in X$, for which $\varphi(x) \neq x$ is finite, we denote by $X_\varphi$ set $X'_\varphi \cup \varphi(X'_\varphi)$, where, as usual, $\varphi(X'_\varphi)$ indicates set $\{\varphi(x) \mid x \in X_\varphi\}$. Evidently, set $X_\varphi$ is finite and function $\varphi$ is a correspondence of set $X_\varphi$ to itself. Therefore, a finite graph with the above-mentioned connected components corresponds to function $\varphi(x)$.

The function $\varphi(x)$ is called the subordination function or function S.

Oriented graphs with numbered vertices corresponding to functions S will be called CH-graphs.

Thus, function S defines a finite CH-graph using the set of vertices $X_\varphi$. To all the other elements $x \in X \setminus X_\varphi$ (whose number may be infinite) correspond isolated vertices, which will be considered "external to the structure" and will not be represented in a figure.

Suppose, now, that set X could be represented as a partition into k minimal subsets $X^i_\varphi$, which makes

$$X_\varphi = \cup X^i_\varphi \quad (\text{for } i \neq j \quad X^i_\varphi \cap X^j_\varphi = \emptyset),\ i = 1, 2, \ldots, k$$

$$\varphi(X^i_\varphi) \subseteq X^i_\varphi$$

correspond to itself.

Then, each of these subsets supplies vertices with the i-th connected component of the CH-graph.

Due to the fact that functions S have a single value, it follows that only a cycle in the i-th connected component corresponds, on graph $\Gamma_\varphi$, to subset $X^{*i}_\varphi \subseteq X^i_\varphi$, such that $\varphi(X^{*i}_\varphi) = X^{*i}_\varphi$. If there is an element $x_n \in X^{*i}_\varphi$ such that $\varphi(x_n) = x_n$, then $X^{*i}_\varphi = \{x_n\}$ and subset $X^i_\varphi$ should not include any other element satisfying this equality. On graph $\Gamma_\varphi$ a tree, whose root is an encircled vertex, corresponds to such a subset.

For $X^{*i}_\varphi$ we will use the term complex cycle of the i-th subset $X^i_\varphi$, and for elements $x \in X_\varphi$ belonging to complex cycles, the term complex elements of graph $\Gamma_\varphi$.

The structure of a CH-graph is univocally determined if, for a given function S, its subset $X^*_\varphi$ and the partition of this subset into cycles $X^{*i}_\varphi$ is found and, further, it is established which of these are isolated.

To determine subset $X^*_\varphi$ it is sufficient to consider a sequence of sets $X_0, X_1, X_2, \ldots, X_k, \ldots$, where $X_0 = X_\varphi$, $X_{i+1} = \varphi(X_i)$, $i = 1, 2, \ldots$

A subset $X^*_\varphi$ is a set $X_k$, therefore $\varphi(X_k) = X_k$. In reality, graph $\Gamma^{k+1}_\varphi$ with vertices taken from set $X_{k+1}$ coincides with graph $\Gamma^k_\varphi$ with vertices taken from set $X_k$, without the hanging vertices. If a CH-graph contains a chain with maximal length k, having no edges common with any of its cycles, then $X^*_k = X^*_\varphi$ and $X^* \subseteq X_\varphi$. To distinguish all cycles $X^{*j}_\varphi$ of subset $X^*_\varphi$, which is the partition of $X^*_\varphi$ into minimal subsets, for which $\varphi(X^{*i}_\varphi) = X^{*j}_\varphi$, requires, in general, a complete analysis of all the subsets in $X^*_\varphi$. To facilitate this search, a procedure can be proposed.

Let be $x \in X^*_\varphi$ and form a chain $X^{*i}_\varphi = \{\varphi^0(x_1), \varphi^1(x_1), \ldots, \varphi^k(x_1)\}$ for which $\varphi^i(x_1) = \varphi(\varphi^{i-1}(x_1))$ and $\varphi^{s+1}(x_1) \in X^{*j}_\varphi$.

This chain always exists and belongs to the j-th connected component.

Now, let $X_2 \in X_\varphi \setminus (X^*_\varphi \cup X^j_\varphi)$ and form, for $x_2$, the same chain $X^{*2}_\varphi$.

This chain will not have intersections with the first one and will represent the second cycle of the CH-graph. By continuing this procedure, the partition of subsets into cycles will be obtained.

Consider, now, the procedure for partitioning the whole set $X\varphi$ of the CH-graph into subsets $X^i_\varphi$.

For $x_1 \in X_\varphi \setminus X^*_\varphi$ we can build a chain $X^1_\varphi = \{x_1, \varphi^1(x_1), \dots, \varphi^s(x_1)\}$ with $\varphi^i(x_1) = (\varphi(\varphi^{i-1}(x_1))$ and $\varphi^{s+1}(x_1) \in X^{*j}_\varphi$. Such a chain always exists and belongs to the j-th connected component. Now, for $x_2 \in X_\varphi \setminus X^*_\varphi \cup X^1_\varphi$ let us construct a chain $X^2_\varphi = \{x_2, \varphi^1(x_2), \dots \varphi^t(x_2)\}$ with $\varphi^{t+1}(x_2) \in X^*_\varphi \cup X^1_\varphi$. If, at the same time, $\varphi^{t+1}(x_2) \in X^{*j}_\varphi \cup X^1_\varphi$ all the elements of this chain also belong to the i-th connected component. Otherwise, $\varphi^{t+1}(x_2) \in X^{*i}_\varphi$ and therefore, all these elements belong to the i-th connected component.

The continuation of such a procedure leads to the partition of set X into its component subsets.

Example 2.1. Let x be a set formed by the first 20 letters of the latin alphabet: X = a, b, c,........,s, t. Using X, two functions $\varphi_1$ and $\varphi_2$ are defined by Table 2.1

| x | a | b | c | d | e | f | g | h | i | j | k | l | m | n | o | p | q | r | s | t |
|---|---|---|---|---|---|---|---|---|---|---|---|---|---|---|---|---|---|---|---|---|
| $\varphi_1(x)$ | b | a | c | d | p | f | a | a | g | j | m | l | b | n | m | a | q | r | m | t |
| $\varphi_2(x)$ | h | s | c | l | l | p | b | f | l | o | b | o | c | s | c | a | o | n | s | l |

Table 2.1

Subsets $x^t$ for these functions are

$$X^1\varphi_1 = \{a, b, c, g, h, i, k, m, o, p, s\}$$

$$X^1\varphi_2 = \{a, b, d, c, f, g, h, i, j, k, l, m, n, o, p, q, r, t\}$$

Functions $\varphi_1$, $\varphi_2$ make these subsets correspond to subsets

$$\varphi_1(X^1\varphi_1) = \{a, b, g, m, p\}$$

$$\varphi_2(X^1\varphi_2) = \{a, b, c, f, h, n, l, o\ p, s\}$$

Subsets $X\varphi$ for both correspondences are

$$X\varphi_1 = X^1\varphi_1 \cup \varphi_1(X^1\varphi_1) = X^1\varphi_1$$

$$X\varphi_2 = X^1\varphi_2 \cup \varphi_2(X^1\varphi_2) = X$$

In order to distinguish subsets $X^*\varphi_1$ and $X^*\varphi_2$, construct the sequence of subsets $X^0\varphi_1, X^1\varphi_1, \dots$ and $X^0\varphi_2, X^1\varphi_2, \dots$ as follows:

$X^0\varphi_1 = X\varphi_1$; $X^1\varphi_1 = \{a, b, g, m, p\}$; $X^2\varphi_1 = \{a, b\}$; $X^3\varphi_1 = X^2\varphi_1$

$X^0\varphi_2 = X\varphi_2$; $X^1\varphi_2 = \{a, b, c, f, h, l, n, o, p, s\}$

$X^2\varphi_1 = \{a, c, f, h, o, p, s\}$; $X^3\varphi_2 = \{a, c, f, h, p, s\}$; $X^4\varphi_2 = X^3\varphi_2$;

thus

$$X^*\varphi_1 = \{a, b\} \; ; \; X^*\varphi_2 = \{a, c, f, h, p, s\}$$

Now, the cycles of subsets $X^*\varphi_1$ and $X^*\varphi_2$ can be constructed. For $X^*\varphi_1$ we find that as $\varphi_1(a) = b$, $\varphi_1(b) = a$, graph $\Gamma\varphi_1$ corresponding to function S $\varphi_1$ contains a single cycle $\{a, b\}$ and therefore it contains only one component connected with a root-cycle consisting of two vertices a and b. For $X^*\varphi_2$, we find that the chain is $\{a, h, f, p\}$ for element $a \in X^*\varphi_2$, whereas for elements c and s the corresponding chains consist only of these same elements. Thus, subset $X^*\varphi_2$ includes three cycles:

$$X^{*1}\varphi_2 = \{a, h, f, p\} \; ; \; X^{*2}\varphi_2 = \{c\} \; ; \; X^{*3}\varphi_2 = \{s\}$$

and therefore graph $\Gamma\varphi_2$ contains three connected components: two trees with roots c and s and a component with a root-cycle $X^{*s}_{\varphi}\ _2 = \{a, h, f, p\}$.

The distinction of different subsets of elements belonging to each component of graph $\Gamma\varphi_2$ can be obtained by numbering the elements not appearing in $X^*\varphi_2$ together with their chains. From this results the following partition of set $X\varphi_2 \setminus X^*\varphi_2$ into the component subsets

$$X^2\varphi_2 \setminus X^{*2}\varphi_2 = \{i, l, o, t, d, e, g, j, m\} \; ;$$

$$X^3\varphi_2 \setminus X^{*3}\varphi_2 = \{x, b, r, n, g\} \; ; \; X^1\varphi_2 \setminus X^{*1}\varphi_2 = \emptyset.$$

The obtained partition proves that $X^{*1}\varphi_2$ is an isolated cycle.

Fig. 2.2 shows graphs $\Gamma\varphi_1$ and $\Gamma\varphi_2$ which correspond to functions S $\varphi_1$, $\varphi_2$

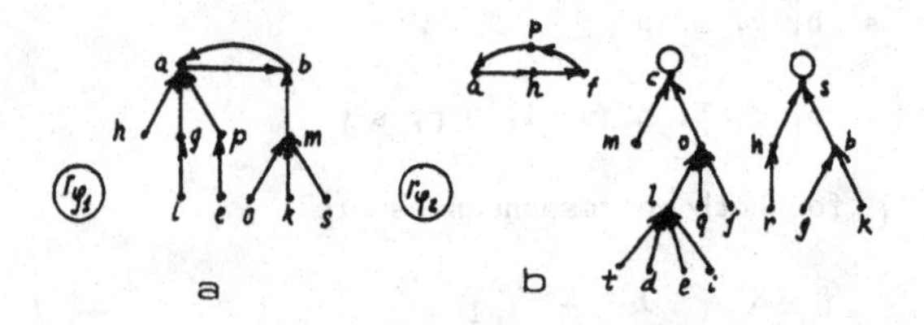


Fig. 2.2

## 2.2 Substitution Operator CH-Graphs Multiplication

Let $\varphi(x)$ be a function S to which a set of vertices $X\varphi$ of graph $\Gamma\varphi$ corresponds, let $\Psi(y)$ be an arbitrary function and let

$$(2.2) \qquad \eta(x) = \Psi(\varphi(x))$$

In this case, we will say that function $\eta(x)$ is obtained by $\varphi(x)$ through the substitution of $\psi$.

Theorem 2.1

In order that set G of CH-graph is closed with respect to substitution 2.2, it is necessary and sufficient that the corresponding function $\psi(y)$, $y \in X$ is a function S'.

Proof

To let $\eta(x)$, defined by 2.2, be a function S, the following conditions are necessary:

a) $\eta(x)$ has a single value

b) $X_\eta$ is fixed

From these conditions and from the finiteness of $X\varphi$ the following properties for function $\psi(y)$ derive:

a') $\psi(y)$ has a single value

b') $X_\psi$ is fixed.

Thus, $\psi(y)$ is a function S. Sufficiency is thus evident.

We will use the term substitution operator $F_\psi$ for the operator performing the transformation of $\varphi(x)$ into $\eta(x)$ according to relation 2.2, with $\psi$ satisfying the condition of theorem 2.1, that is, with $\psi$ as a function S. In these terms, theorem 2.1 means that despite the application of the substitution operator to a CH-graph, this remains in the class of CH-graphs. The substitution operator is a local operator: the subordination $\eta(x)$ of vertex x in the graph, produced by the substitution, depends only on the subordination $\varphi(x)$ in the initial graph and does not depend on the subordination structure of other vertices of the graph.

According to theorem 2.1, a CH-graph $\Gamma_\psi$ defined using X, can be made to correspond to each substitution $\psi(\varphi)$. This graph will be called substitution graph $\Gamma_\psi$ or, in short, substitution.

In the language of graphs, a multiplication on graphs[1] can be made to correspond to an operation $\eta(x) = \psi(\varphi(x))$

$$(2.3) \qquad \Gamma_\eta = \Gamma_\varphi * \Gamma_\psi$$

This multiplication has the following properties:

1) it makes the set of vertices of graphs $\Gamma_\varphi$ and $\Gamma_\psi$ to correspond to themselves.

$$X_\eta \subseteq X\varphi \cup X\varphi$$

In fact, if $x \in X \setminus (X\psi \cup X\varphi)$ then $\varphi(x) = \psi(x) = x$, that is $\eta(x) = x$; from this, it follows that $X\psi \cup X\varphi$ is made to correspond to itself. This property assures that the appearance of new vertices on graph $\Gamma_\eta$ is possible only when $X\psi \nsubseteq X\varphi$, whereas for $X_\psi \subseteq X_\varphi$ only the reconstruction of the graph, but not its growth, is possible.

2) Associativity.

If $\Gamma_\eta = \Gamma_\varphi * \Gamma\varphi_1 * \Gamma\varphi_2$ and $\Gamma_\psi = \Gamma\varphi_1 * \Gamma\varphi_2$

then $\Gamma_\eta = \Gamma_\varphi * \Gamma_\psi$

The validity of this can be easily demonstrated through the sequential substitution of functions S in graphs $\Gamma_{\varphi_1}$, $\Gamma_{\varphi_2}$ and $\Gamma_{\psi}$

3) Conservation of complexes.

For a given graph $\Gamma_{\varphi}$ a complex is any subset $X_k \subseteq X$ for which $\varphi(x) = y =$ constant, $x \in X_k$. Thus a complex is formed by a subset of vertices in $\Gamma_{\varphi}$ subordinated to the same vertex which is their common "leader".

If $\Gamma_{\eta} = \Gamma_{\varphi} * \Gamma_{\psi}$ and $X_k$ is a particular complex of graph $\Gamma_{\varphi}$, then subset $X_k$ of graph $\Gamma_{\eta}$ is also a complex.

This property guarantees the "non destruction" of complexes, but it does not exclude the union of some new vertices or integer complexes to already existing complexes.

4) Conservation of non-intersecting components.

If one of the multiplicands in product $\Gamma_{\eta} = \Gamma_{\varphi} * \Gamma_{\psi}$ contains components $\Gamma_{\varphi_i}$ or $\Gamma_{\psi_i}$ for whose correspondent subsets $X^i_{\psi}$ and $X^i_{\varphi}$ conditions $X^i_{\psi} \cap X_{\varphi} = \emptyset$, $X^i_{\varphi} \cap X_{\varphi} = \emptyset$ are satisfied, then these components remain unaltered in graph $\Gamma_{\eta}$.

From the point of view of the structural properties of set G in all CH-graphs with vertices in X, the following characteristic subsets can be distinguished: subset T consisting of all the trees and forests; subset C, including cycles and their combinations; subset CT, containing components of mixed type: trees and cycles, cycles with hanging trees, and so on (Fig. 2.3)

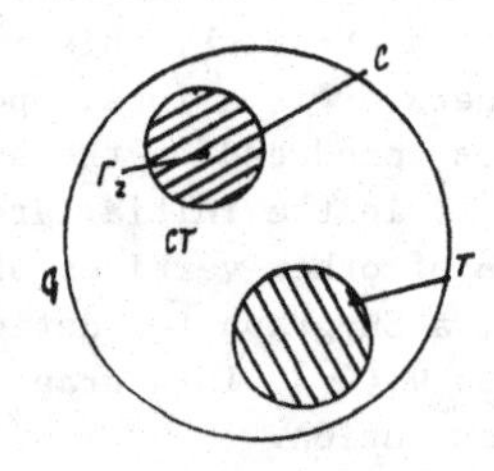


Fig. 2.3

Ficticius graph $\Gamma_z$ with function S $\varphi(x) = x$, $x \in X$ belongs to subset C. If we consider the encircled isolated vertices, then $\Gamma_z$ represents the set (generally finite) of encircled isolated vertices.

As these isolated vertices have been initially excluded from the structure of CH-graphs, the structure of graph $\Gamma_z$ is empty and this graph represents the unity for operation * on CH-graphs, because, evidently,

$$\Gamma * \Gamma_z = \Gamma_z * \Gamma = \Gamma \quad \text{for any} \quad \Gamma \in G$$

5) If $\Gamma = \Gamma_1 * \Gamma_2$ and at least one of the graphs $\Gamma_1$ or $\Gamma_2$ does not belong to subset C, then $\Gamma \in C$. The significance of this property is that for multiplication * a graph can be taken from subset C, but the inverse translation cannot be obtained. Thus, from this property and from the fact that $\Gamma_z \in C$ it follows that, with respect to operation * with unity $\Gamma_z$, graphs T and CT do not have their inverse in G.

Operation * is not generally commutative.

From this property it also follows that operation * generates a non-commutative semi-group with identity on the set G of all CH-graphs. As graphs C are nothing else than the well-known algebraic substitutions, and operation * is their multiplication, this operation generates a group with the same identity on C.

Next theorems 2.2 and 2.3 refer to the problem of the representation of CH-graphs in the form of a product *.

Theorem 2.2
Two graphs $\Gamma_\eta$ and $\Gamma_\varphi$ having been assigned, so that in $\Gamma_\eta$ all the complexes of $\Gamma_\varphi$ are present, there is a unique graph $\Gamma_\psi$ for which $\Gamma_\eta = \Gamma_\varphi * \Gamma_\psi$.

Proof
According to the properties of conservation of complexes , all the complexes of graph $\Gamma_\varphi$ remain in $\Gamma_\eta$ for $\Gamma_\varphi * \Gamma_\psi = \Gamma_\psi$. To prove that graph $\Gamma_\psi$ always exists, it is sufficient to construct its generating function $\psi(x)$, and to demonstrate that it is a function S, and that $\psi(\varphi(x)) \equiv \eta(x)$.

This function $\psi$ is defined by conditions:

a. for $x \in X_\eta \setminus X_\varphi$ (x) = (x);
b. if $x \in X_\eta \cap X_\varphi$ and z exists for which $\varphi(z) = x$, then $\psi(x) = \eta(x)$;
c. if $x \in X_\varphi \setminus X_\eta$ and z exists for which $\varphi(z) = x$, then $\psi(x) = z$;
d. otherwise, $\psi(x) = x$.

Evidently, $\psi(x) = x$ for $x \notin X_\varphi \cup X_\eta$ and $\psi(x)$is univocally defined for every x. It follows that $\psi(x)$ is a function S.

We show that $\psi(\varphi(x)) = \eta(x)$. Let $x \notin X \setminus X_\varphi$ and $\eta(x) = y$. Then $\varphi(x) = x$ and $\psi(x) = \eta(x) = y$ and therefore $\psi(\varphi(x) = \psi(x) =$ $= y = \eta(x)$. In addition, let $x \in X_\varphi \setminus X_\eta$ and $\varphi(x) = y$. Then $y \in X_\varphi$ and x is the unique vertex for which $\varphi(x) = y$ (otherwise the conservation property of complexes would not be respected), and thus from the definition of $\psi$ it would result $\psi(y) = x$. From this, $\psi(\varphi(x)) = \psi(y) =$ $= x$ $\eta(x)$. Now, let $x \in X_\varphi \cap X_\eta$ and $\varphi(x) = y$, $\eta(x) = z$. Then $y \in X_\varphi$, $\psi(y) = z$ and for every vertex v for which $\varphi(v) = y$ is $\eta(x) =$ $= z$ (due to the conservation property of complexes).So, $\psi(\varphi(x)) = \psi(y) =$ $= z = \eta(x)$. Otherwise, evidently, $\psi(x) = \varphi(x) = \eta(x) = x$.

Example 2.2 In Figs. 2.4a and b, graphs $\Gamma_\varphi$ and $\Gamma_\eta$ , satisfying the conditions of theorem 2.2, have been drawn. Table 2.2 is the table of function S $\varphi(x)$ defined according to conditions a) - d)

| x | a | b | c | d | e | f | g | h | i | j | k | l |
|---|---|---|---|---|---|---|---|---|---|---|---|---|
| $\varphi(x)$ | l | c | k | d | j | f | g | j | j | i | i | l |

Table 2.2

Fig. 2.4c shows the graph corresponding to this function.
It is easy to verify that $\Gamma_\eta = \Gamma_\varphi * \Gamma_\psi$

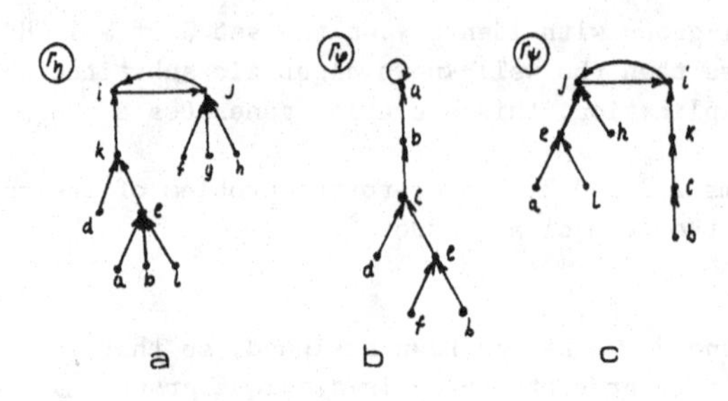


Fig. 2.4

Theorem 2.3

a) Any graph T $\Gamma_\eta$ with no less than three vertices can be represented in the form of a product $\Gamma_\eta = \Gamma_\varphi * \Gamma_\psi$ where $\Gamma_\varphi$ and $\Gamma_\psi$ are graph T, such that subset $X_\varphi \cup X_\psi$ contains a single vertex - tree's root $\Gamma_\varphi$ or $\Gamma_\psi$ (Fig. 2.5a).

b) Any graph CT can be represented in the form of a commutative product $\Gamma_\eta = \Gamma_\varphi * \Gamma_\psi$ where one of the multiplicands is a graph C and the other is a graph T (Fig. 2.5b).

c) Any graph CT can be represented in the form of the product (in general non-commutative) of two graphs T (Figs. 2.5c and d)

Proof

Parts a) and b) of the theorem are evident. To demonstrate part c), we start with two simple cases of graphs CT: a cycle with hanging trees (Fig. 2.5c) and a graph CT with two components consisting of an isolated cycle and a tree (Fig. 2.5d). Let x be a vertex of some of the trees hanging from the cycle, and, in graph $\Gamma_\eta$, let y be a vertex to which x is subordinated (i. e. $y = \eta(x)$).

Graph $\Gamma_\varphi$ in $\Gamma_\eta = \Gamma_\varphi * \Gamma_\psi$ is formed by a chain with root x to which the vertex of cycle z, for which $\eta(z) = y$, is directly subordinated, and the remaining part of the chain is built with a fragment of a cycle of arbitrary length which starts from 2. Let U be the last element (which is the hanging element of chain $\Gamma_\varphi$) of the fragment, and let us take as $\Gamma_\psi$ a chain with root U, formed from the remaining fragment of the cycle (from U to y), from vertex x hanging from y and from the remaining fragments of graph $\Gamma_\eta$. In Fig. 2.5c, graphs $\Gamma_1$ and $\Gamma_2$, constructed in such a manner, are shown. It is easy to verify that their product gives $\Gamma$.

Let us consider now Fig. 2.5d. Cut the cycle at an arbitrary point and connect the chain so obtained to any hanging vertex of the tree, for example, to vertex x, and mark with y the external disconnected cycle. The graph so obtained will be similar to $\Gamma_1$. In the case of graph $\Gamma_2$ we will choose the chain contained between two vertices: root y and the part hanging from vertex x. The product of $\Gamma_1$ and $\Gamma_2$ will give us initial graph CT $\Gamma$. Fig. 2.5d shows graphs $\Gamma_1$ and $\Gamma_2$ constructed in such a way. It is easy to verify that their product now gives graph $\Gamma$.

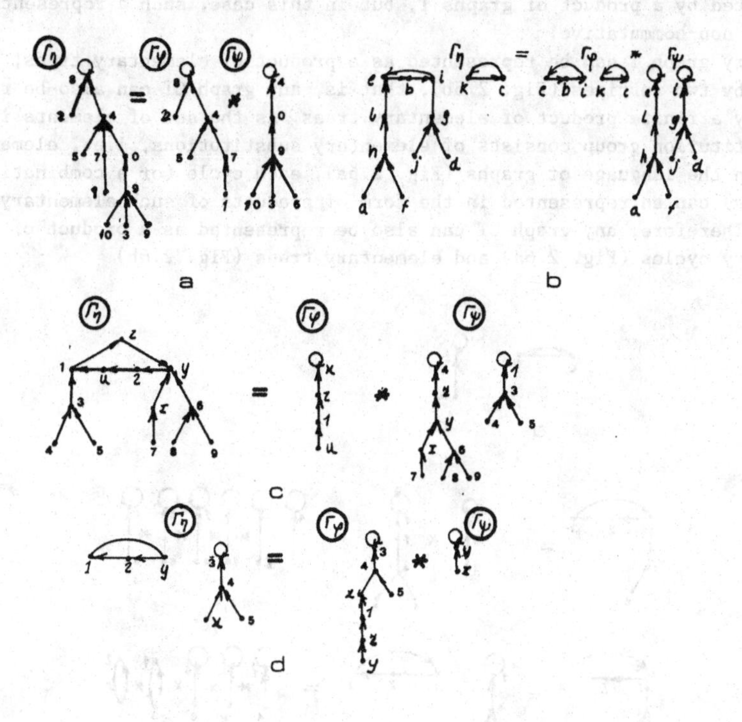


Fig. 2.5

Generally, a graph CT may contain arbitrary combinations of a - c type components (see Fig. 2.1). Applying successively the afore-mentioned techniques to construct graphs $\Gamma_\varphi$ and $\Gamma_\psi$ and considering the associative property of operation *, we will find that an arbitrary graph CT can be represented by a product of graphs T, but in this case, such a representation would be non-commutative.

Every graph T can be represented as a product of elementary trees, formed by two vertices (Fig. 2.5b), that is, any graph CT can also be represented by a finite product of elementary trees. As the set of elements in the substitution group consists of elementary substitutions, i.e., elementary cycles in the language of graphs (Fig. 2.6a), each cycle (or a combination of cycles) can be represented in the form of products of such elementary cycles. Therefore, any graph CT can also be represented as a product of elementary cycles (Fig. 2.6a) and elementary trees (Fig. 2.6b)

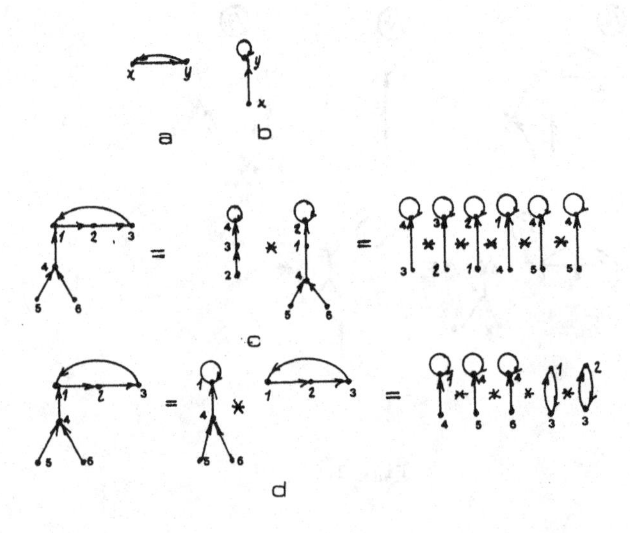


Fig. 2.6

For this reason, a combination of elementary cycles and trees forms the system of elements in semi-group G of CH-graphs. In Fig. 2.5c and d, two different division procedures of graph CT into elements are shown.

## 2.3 Fundamental Equation and General Properties of Graph - Trajectories

Let x be a finite or infinite set of elements; G the set of all CH-graphs, whose vertices are elements $x \in X$; $\varphi(x)$ a function S representing initial CH-graph $\Gamma \in G$; $X_\varphi$ the set of its vertices; $F_\psi$ a substitution operator to which graph $\Gamma_\psi \in G$ corresponds; $\psi(x)$ and $X_\psi$ function S and the set of vertices of $\Gamma_\psi$, respectively.

The equation connecting the state of variable graph $\Gamma_\eta(t)$ with $\Gamma_\eta(t-1)$ can be written in the form

$$\Gamma_\eta(t) = F_\psi \Gamma_\eta(t-1); \qquad \Gamma_\eta(0) = \Gamma \tag{2.4}$$

or as a product of CH-graphs

(2.5) $\Gamma_\eta(t) = \Gamma_\eta(t-1) * \Gamma_\eta(0) = \Gamma_\varphi$

Equation 2.4 or 2.5 makes it possible to determine the graph trajectory which is a sequence of CH-graphs

(2.6) $\Gamma_\eta(0), \quad \Gamma_\eta(1), \quad \Gamma_\eta(2) \ldots\ldots$

for a given initial graph.

If we take into consideration the associative property of multiplication, equation 2.5 may be re-written as

(2.7) $\Gamma_\eta(t) = \Gamma_\varphi * \Gamma^t_\psi$ where $\Gamma^t_\psi = \Gamma_\psi * \Gamma_\psi * \ldots * \Gamma_\psi$

Now some definitions are necessary.

We will call a graph, for which

(2.8) $\tilde{\Gamma}_\eta(t+T) = \tilde{\Gamma}_\eta(t)$

a periodical solution of equation 2.7, with period T.

When T = 1, the solution is said to be stationary, and it is denoted by $\bar{\Gamma}_\eta$

(2.9) $\bar{\Gamma}_\eta * \Gamma_* = \Gamma_\eta$

The maximal succession $\Gamma_\eta(0), \Gamma_\eta(1), \ldots, \Gamma_\eta(k-1)$ containing fragments of a stationary or periodical solution represents the non-stationary part of the trajectory of equation 2.5.

It has been previously proved that, if $\Gamma_\eta = \Gamma_\varphi * \Gamma_\psi$, then $X_\eta \subseteq X_\varphi \cup X_\psi$ and therefore $X_\eta(t) \subseteq X_\varphi \cup X_\psi$ for every t. As long as $X_\varphi$ and $X_\psi$ are finite and constant, only a finite number of different graphs belongs to sequence 2.6, that is, the sequence contains a finite, non-stationary part, after which it becomes periodical. This is true for any substitution operator and for any initial graph $\Gamma_\varphi$.

Now let $\Gamma_{\varphi 1}, \Gamma_{\varphi 2}, \ldots, \Gamma_{\varphi k}$ be the connected components of the initial graph $\Gamma_\varphi$ ; $X^1_\varphi, X^2_\varphi, \ldots, X^k_\varphi$ the correspondent sets of vertices; $X^{*1}_\varphi, X^{*2}_\varphi, \ldots, X^{*k}_\varphi$ the cycles; and $X^*_\varphi = \bigcup_{j=1}^{k} X^{*j}_\varphi$ the set of complex vertices of integer graph $\Gamma_\varphi$. Correspondingly, for substitution graph $\Gamma_\psi$, let $\Gamma_{\psi 1}, \Gamma_{\psi 2}, \ldots \Gamma_{\psi m}$ be the connected components; $X^1_\psi, X^2_\psi, \ldots X^m_\psi$ the component sets; $X^{*1}_\psi, X^{*2}_\psi, \ldots, X^{*m}_\psi$ the cycles and $X^*_\psi$ the complex set of graph $\Gamma_\psi$. We will call index $S^{x*}_x$ of a vertex $x \in X^j_\psi$ with respect to complex element $x^* \in X^{*j}_\psi$ the length of the line from x to $x^*$, and module, the length $l^j_\psi$ of the correspondent cycle $X^{*j}_\psi$. A vertex is subordinated to a cvcle, if it is subordinated to one of its elements; although it is not always necessarily subordinated to the same element. The behaviour of separate vertices $x \in X_\varphi \cup X$ along graph trajectory (2.6) is determined by the following lemma.

Lemma

Along the trajectory of graph 2.6 of equation 2.5, the following statements are valid:

1) Any vertex, for which $X_2 = \varphi(x_1) \in X^j_\psi$ will be subordinated to cycle $X^{*j}_\psi$, starting from $t = t^*$;
2) vertices $x_s \in X_\varphi \cup X_\psi$ belong to the same complex of a periodical or stationary solution, if, and only if, they are subordinated to elements $x_p = \varphi(x_s)$ so that all the x's belong to the same component graph $\Gamma_\psi$ and their indices $S^{xp}_{x^*}$ with respect to an arbitrary but fixed vertex $x^*$, are equal for all x's. All vertices $x_s$ of graph $\Gamma_\varphi$, for which $x_p = \varphi(x_s) \notin X_\psi$ keep their subordination at each step.

Proof

It is easy to verify that for any graph $\Gamma_\psi$, there is a time t* such that for $t > t^*$ any complex of vertices $x \in X^j_\psi$ of graph $\Gamma^t_\psi$ must be subordinated to one of vertices $x^* \in X^{*j}_\psi$.

It is also evident that $x_1, x_2 \in X_\varphi$ depend on the same complex if, and only if, there is a t for which $\psi^t(\varphi(x_1)) = \psi^t(\varphi(x_2))$, but, at the same time, the equality $\psi^t(x_3) = \psi^t(x_4)$ is true for $x_3, x_4 \in X_\psi$ if, and only if, indices of $x_3$ and $x_4$, with respect to an arbitrary, complex common element c, are equal.

If for a component $\Gamma_{\varphi_i}$ of initial graph $\Gamma_\varphi$, the relation $X^{*i} \cap X_\psi = \emptyset$ is true, then, according to the lemma, the cycle along trajectory 2.6 is invariable and the subordination of all the vertices subordinated to the cycle remains unaltered. Such a component does not structurally vary along trajectory 2.6, even if the set of its vertices does, and it will be called a trivial component of initial graph $\Gamma_\varphi$ and graph $\Gamma_\eta(t)$. Component $\Gamma_{\varphi_i}$ for which $X^i_\varphi \cap X_\psi = \emptyset$ and which remains completely unchanged along trajectory 2.6, is called absolutely trivial. The remaining components of $\Gamma_\eta(t)$ are not trivial. When, for initial graph $\Gamma_\varphi$, relation $X_\varphi \cap X_\psi = \emptyset$ is valid, graph trajectory 2.6 is of the form

$$\Gamma_\eta(0) = \Gamma_\varphi ; \quad \Gamma_\eta(1) = \Gamma_\varphi \cup \Gamma_\psi ; \quad \Gamma_\eta(2) = \Gamma_\varphi \cup \Gamma^2_\psi \; \ldots\ldots$$

and graph $\Gamma_\varphi$ "accompanies" the trajectory $\Gamma_\psi, \Gamma^2_\psi, \Gamma^3_\psi, \ldots\ldots$

Example 2.3 X is a set of integers and function S of the substitution operator is given in taple 2.3 (for all the x's not mentioned in table 2.3, $\varphi(x) \equiv x$)

| x | 1 | 2 | 3 | 4 | 5 | 6 | 7 | 8 | 9 | 10 | 11 | 12 | 13 | 14 | 15 |
|---|---|---|---|---|---|---|---|---|---|---|---|---|---|---|---|
| $\psi(x)$ | 10 | 1 | 8 | 11 | 2 | 2 | 10 | 9 | 8 | 7 | 8 | 3 | 7 | 11 | 12 |

Table 2.3

Substitution graph $\Gamma_\psi$ (Fig. 2.7a) has two components ($\psi_1$ and $\psi_2$) with component subsets $X^1_\psi = \{1, 2, 5, 6, 7, 10, 13\}$ and $X^2_\psi = \{3, 4, 8, 9, 11, 12, 14, 15\}$. The initial graph, which has also two components, is shown in Fig. 2.7b). Fig. 2.7c shows the trajectory of equation 2.5 for $t = 1 \div 5$. As shown in Fig. 2.7c, graph $\Gamma_\eta(t)$ is periodical with period $T = 2$, which strts from $t = 3$. $\Gamma_{\varphi_1}$ is a trivial component, $\Gamma_{\varphi_2}$ an absolutely trivial component.

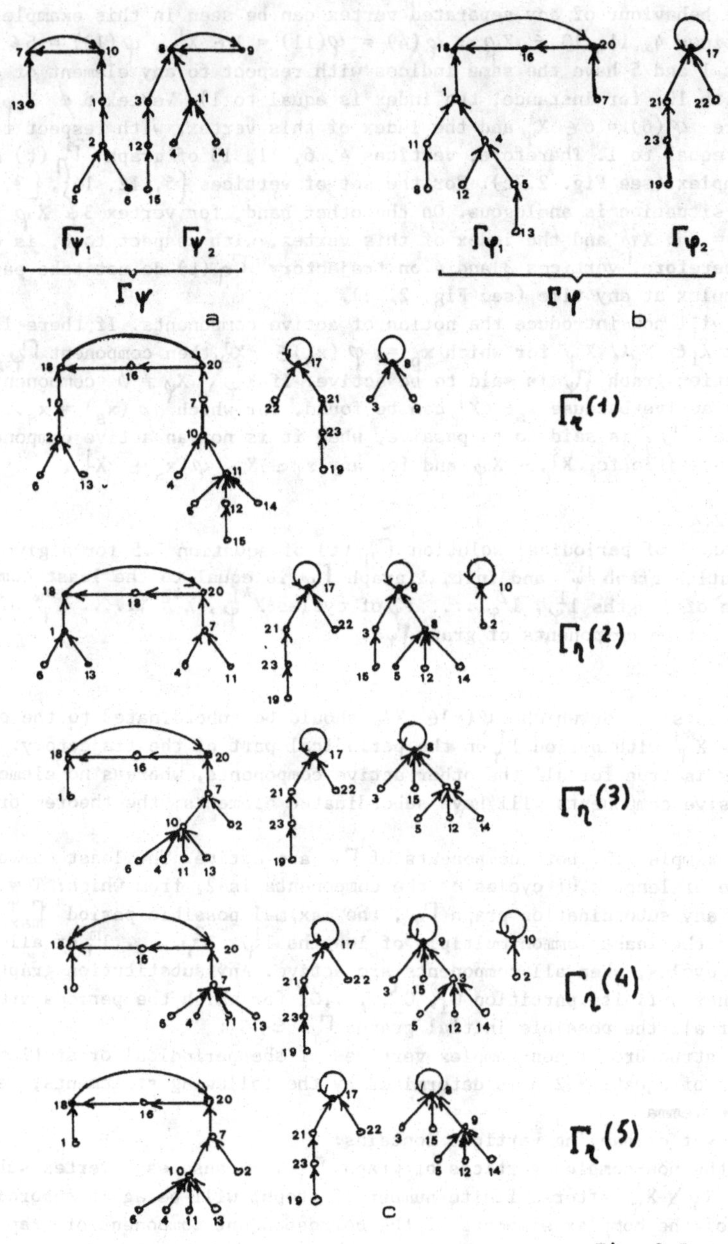


Fig. 2.7

The behaviour of any separated vertex can be seen in this example. For vertices $4, 11, 13 \in X_\varphi$ : $\varphi(4) = \varphi(11) = 1 \in X^1_\psi$, $\varphi(13) = 5 \in X^1_\psi$, vertices 1 and 5 have the same indices with respect to any element of component $\Gamma_{\psi\,1}$(with 10, for instance, the index is equal to 1). Vertex $6 \notin X_\varphi$, and therefore $\varphi(6) = 6 \in X^1_\psi$ and the index of this vertex, with respect to 10, is also equal to 1. Therefore, vertices 4, 6, 11, 13 of graph $\tilde{\Gamma}_\eta(t)$ make up a complex (see Fig. 2.7c). For the set of vertices $\{5, 12, 14\}$, $\{3, 9, 15\}$ the situation is analogous. On the other hand, for vertex $3 \notin X_\varphi$ : : $\varphi(3) = 3 \in X^2_\varphi$ and the index of this vertex, with respect to 8, is equal to 1. Therefore, vertices 3 and 4 on trajectory $\Gamma_\eta(t)$ do not take part in a complex at any time (see Fig. 2.7c).

We will now introduce the notion of active components. If there is a vertex $X_1 \in X_\varphi \cup X_\psi$ for which $x_p = \varphi(x_s) \in X^j_\psi$ then component $\Gamma_{\psi\,i}$ of substitution graph $\Gamma_\psi$ is said to be active. If $X^j_\psi \setminus X_\varphi \neq \emptyset$ component $\Gamma_{\psi\,j}$ is active because $x_s \in X^j_\psi$ can be found, for which $\varphi(x_s) = x_s$. Component $\Gamma_{\psi\,j}$ is said to be passive, when it is not an active component. This is possible for $X^j_\psi \subseteq X_\varphi$ and for any $x_1 \in X_\varphi$, $\varphi(x_s) \in X^j_\psi$.

Theorem 2.4

The period T of periodical solution $\tilde{\Gamma}_\eta(t)$ of equation 2.5 for a given substitution graph $\Gamma_\psi$ and initial graph $\Gamma_\varphi$ is equal to the least common multiple of lengths $1^1_\psi, 1^2_\psi, \ldots\ldots 1^w_\psi$ of cycles $X^{*1}_\psi, X^{*2}_\psi, \ldots\ldots, X^{*w}_\psi$ of all the active components of graph $\Gamma_\psi$.

Proof

All elements x, for which $\varphi(x) \in X^j_\psi$ should be subordinated to the elements of cycle $X^{*j}_\psi$ with period $1^j_\psi$ on the periodical part of the trajectory. Similarly, this is true for all the other active components, whereas no element of the passive components will have subordinated elements; the theorem proves this.

In example 2.3, both components of $\Gamma_\psi$ are active. The least common multiple of lengths of cycles of the components is 2, from which: T = 2.

For any subordination graph $\Gamma_\psi$, the maximal possible period $\Gamma^\psi_{max}$ is equal to the least common multiple of lengths $1^1_\psi, 1^2_\psi, \ldots\ldots 1^m_\psi$ of all complex cycles, when all components are active. Any substitution graph $\Gamma_\psi \in G$ determines a finite partition $G^\psi_1, G^\psi_2, \ldots, G^\psi_r$ for which the periods are the same for all the possible initial graphs $\Gamma_\varphi \in G_j$.

The structure of non-complex vertices of the periodical or stationary solution of equation 2.5 is determined by the following statements, deriving from the lemma.

The set of hanging vertices contains:

a. all the non-complex vertices of graph $\Gamma_\psi$, because each vertex subordinated to $x \in X_\psi \setminus X^*_\psi$, after a finite number of steps, will be again subordinated to one of the complex elements of the correspondent component of graph $\Gamma_\psi$;

b. all the hanging vertices x of graph $\Gamma_\varphi$ are such that $x \in X_\psi$;

c. all complex elements of passive components of substitution graph $\Gamma_\psi$ have already lost, at the first step, all the vertices which were directly subordinated to them and successively, they do not receive any new element;

d. some complex elements of the active components of graph $\Gamma_\psi$ can be periodically hanging due to subordinated vertices being cyclically transferred from one complex vertex to another.

Non-hanging vertices $x \in \Gamma_\varphi$ for which relation $x \notin X_\psi$ is true, are non-hanging on trajectory 2.6 at every instant.

We have previously shown that $X_\eta(t) \subseteq X_\varphi \cup X_\psi$.

This inclusion means that some vertices of graphs $\Gamma_\varphi$ and $\Gamma_\psi$ are about "to fall" from the structure of graphs $\tilde{\Gamma}_\eta(t)$.

From the afore given description of sets of hanging and non-hanging vertices, it follows that only complex vertices of graph $\Gamma_\psi$ can fall.

Example 2.4. Set x contains the letters of the Slavic and Latin alphabets; function S is determined in this domain by table 2.4 and $\psi(x) \equiv x$ for values $x \in X$ not contained in the table. Substitution graph $\Gamma_\psi$ (Fig. 2.8a) contains four components $\Gamma_{\psi 1}, \ldots\ldots, \Gamma_{\psi 4}$; the initial graph is shown in Fig. 2.8b.

| x | а | б | в | г | д | е | ж | з | и | к | л | м | н | о | п | р | с | т | у | ф | х | ц | ч |
|---|---|---|---|---|---|---|---|---|---|---|---|---|---|---|---|---|---|---|---|---|---|---|---|
| φ(x) | к | а | з | и | б | б | к | и | з | ж | з | в | ж | и | м | э | у | ф | ф | э | с | ж | ч |

Table 2.4

Component $\Gamma$ is trivial because $X^{*1}_{\varphi} \cap X_\psi = \emptyset$.

On the other hand, any vertex $x \in X^2_\psi$ is hanging in graph $\Gamma_\varphi$ and, therefore, component $\Gamma_{\psi 2}$ is passive, whereas the three remaining components of graph $\Gamma_\psi$ are active. The least common multiple of the lengths of collective cycles of active components $\Gamma_{\psi 1}$, $\Gamma_{\psi 3}$ and $\Gamma_{\psi 4}$ is equal to 6, therefore, period T = 6. Fig. 2.8c shows a fragment of the trajectory of fundamental equation (2.5) for $t = 1 \div 9$. The trajectory contains a periodical part with T = 6, which begins with t = 3. Component $\Gamma_{\varphi 1}$ conserves its own structure along the periodical trajectory, even if the vertices of these components vary.

Consider, in this example, the structure of the set of non-complex vertices.

Vertices {л м у ж} $\in X_\psi \setminus X^*_\psi$ and therefore they are hanging for any t on the periodical part of the trajectory.

Hanging vertices д , г of graph $\Gamma_\varphi$ do not belong to the complex subset of graph $\Gamma_\psi$ and, therefore, they are also hanging on the periodical part of the trajectory. So too for complex elements з and и of passive component $\Gamma_{\psi 2}$ of $\Gamma_\psi$.

## 2.4 Envelope of complex subsets of the periodical solution

In the periodical solution $\tilde{\Gamma}_\eta(t)$, complex invariable cycles correspond to trivial or absolutely trivial components of initial graph $\Gamma_\varphi$, and these cycles coincide with the cycles of these components. Later, we will investigate only the properties of the part of the complex subset of graph

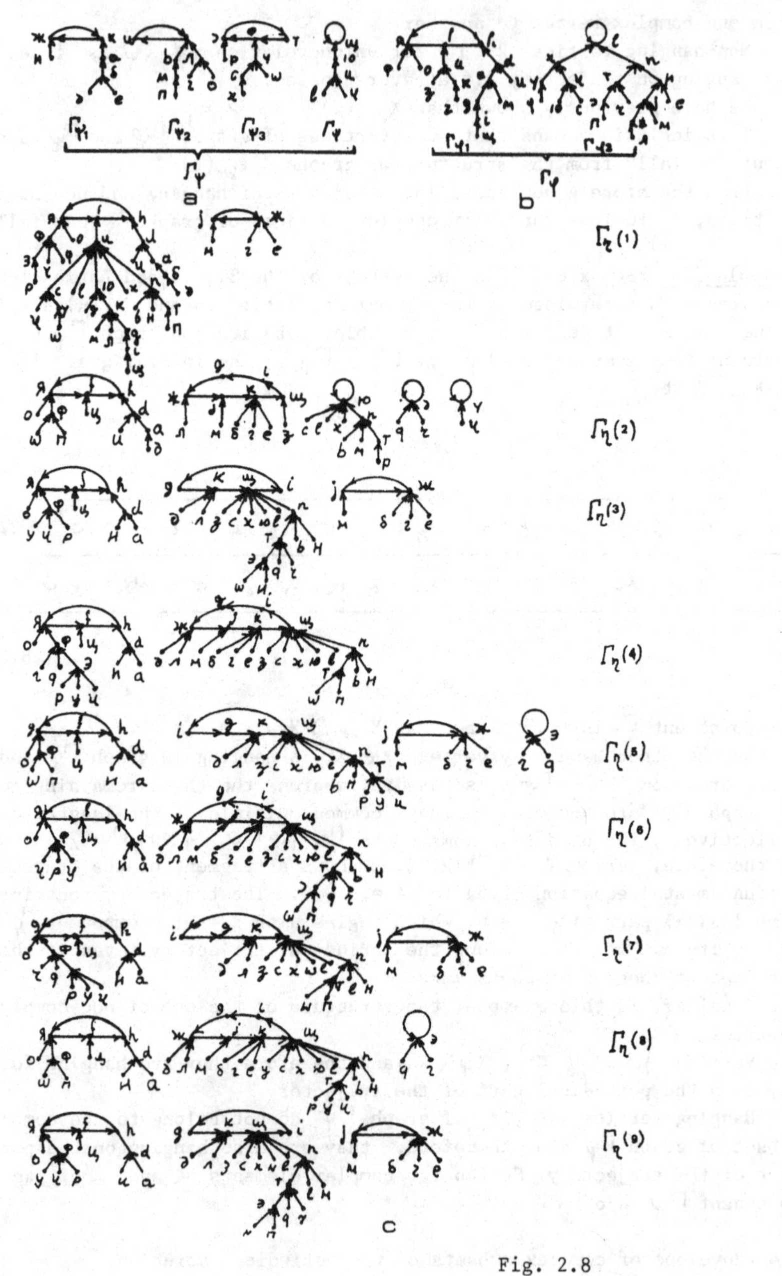
Γψ1
Γψ2
Γψ3
Γψ
Γψ
a
Γφ1
Γφ2
Γφ
b
Γη(1)
Γη(2)
Γη(3)
Γη(4)
Γη(5)
Γη(6)
Γη(7)
Γη(8)
Γη(9)
c

Fig. 2.8

$\tilde{\Gamma}_\eta(t)$ which does not include these invariable cycles, and which will be denoted by $X^*_\eta(t)$.

Subset $X^*_\eta(t)$ of graph $\tilde{\Gamma}_\eta(t)$ can, in general, vary along the trajectory, therefore it is useful to introduce the notion of complex envelope of solution $\tilde{\Gamma}_\eta(t)$. A complex envelope $E_T(X^*_\eta(t)) \subseteq X_\varphi \cap X_\psi$ forms a set for which:

1) for any $t_\eta$ $X^*_\eta(t) \subseteq E_T(X^*_\eta(t))$
2) for any $x \in E_T(X^*_\eta(t))$ there is a $t \in [0, T]$ for which $x \in X^*_\eta(t)$.

In other words, a complex envelope is formed by the set of all the elements $x \in X_\varphi \cup X_\psi$ which are complex at least once in a period. Evidently, $E_T(X^*_\eta(t)) \neq \emptyset$ if, and only if, the components of graph $\Gamma_\varphi$ are trivial or absolutely trivial. In example 2.4, the complex envelope is

$$E_T(X^*_\eta(t)) = \{ \text{ж} \;\; \text{к} \;\; \text{щ} \;\; g \;\; i \;\; \iota \;\; \ni \}$$

We now introduce the notion of p(x) chain of element $x \in X$ and its projection. A p(x) chain is a finite sequence $\varphi^0(x), \varphi^1(x), \dots \varphi^h(x)$ such that $\varphi^0(x) = x$, $\varphi^i(x) = \varphi(\varphi^{i-1}(x))$, $\varphi^{h+1}(x) \in X_\psi$ and $\varphi^i(x) \notin X_\psi$ for $i < h$.

In general, it is possible that a p(x) chain does not exist, for instance, for any $x \notin X_\varphi \cup X_\psi$. If the chain exists and has a length h, we will say that element x is projected on component $\Gamma_{\psi_5}$ of graph $\Gamma_\psi$ and that $\pi(x) = \varphi^{h+1}(x)$ is its projection.

Passive components of graph $\Gamma_\psi$ do not contain any projection.

To determine the content of the complex envelope of graph $\tilde{\Gamma}_\eta(t)$, we show an auxiliary numbered digraph $\Gamma''_{\varphi\psi}$, not belonging to the class of CH-graphs, whose vertices will be integers and the indices of its sides, elements of x.

To do this, we assign arbitrary numbers to components of graph $\Gamma_\psi$ and make a vertex of graph $\Gamma'_{\varphi\psi}$ with the same index correspond to each component $\Gamma_{\psi_j}$ with index j.

Further, for any complex element $x \in X^{*j}_\psi$ of $\Gamma_\psi$ projecting itself on component $\Gamma_{\psi_i}$, we introduce a line on graph $\Gamma'_{\varphi\psi}$ which goes from vertex j to vertex i, and to this we assign index x. In this case we say that element x has a side in graph $\Gamma'_{\varphi\psi}$. The number of sides coming from vertex j, including loops, does not exceed the length of collective cycle $X^{*j}_\psi$. We now define a subgraph $\Gamma_{\varphi\psi}$ of graph $\Gamma'_{\varphi\psi}$.

For this reason, we consider a cycle of graph $\Gamma'_{\varphi\psi}$, containing vertices $j_1, j_2, \dots, j_k$ and sides $x_k, x_1, \dots x_{k-1}$ (side $x_1 = (j_i, j_{i+1}), \dots$ side $x_k = (j_k, j_1)$). We assume that $j_p$ and $j_s$ are two vertices of this cycle, for which lengths $l^{j_p}_\psi$ and $l^{j_s}_\psi$ of the corresponding[2] complex cycles of graph $\Gamma_\psi$ are multiple.

When the the projection index $\pi(x_{p-1})$, with respect to complex vertex $x_p$, is equal to the projection index $\pi(x_{s-1})$, with respect to $x_s$, we say that vertices $j_p$ and $j_s$ satisfy the equality condition of indices. If any pair of vertices belonging to this cycle and having multiple lengths satisfies the equality condition of indices, we say that, on the whole, the cycle satisfies this condition. First of all, we will delete from graph $\Gamma'_{\varphi\psi}$ all the sides not entering the oriented cycles of the graph (the encircled vertex has here the same value as a cycle). From the remainig sides, we will first delete those not entering the cycles of graph $\Gamma'_{\varphi\psi}$ and which satisfy the equality condition of indices, and then all vertices which become isolated due to this deletion. The subgraph of the resulting graph $\Gamma'_{\varphi\psi}$ is denoted by $\Gamma_{\varphi\psi}$.

Theorem 2.5

The collective envelope $E_T(x^*\eta(t))$ consists exclusively of all those vertices $x \in X^*_\psi$ which have sides in graph $\Gamma_{\varphi\psi}$ and of the elements of their p(x) chain.

Proof

Let $N = \{j_1, j_2, \ldots, j_k\}$ be a set of vertices of a cycle of graph $\Gamma_{\varphi\psi}$; $X = \{x_1, x_2, \ldots, x_k\}$ the set of sides of this cycle; $L = \{1^1_\psi, \ldots, 1^k_\psi\}$ the set of the lengths of vertices $j_i \in N$. To begin with, we assume that L does not contain pairs of multiple numbers and $p(x_i) = x_i$ for any $x_i \in X$. We will consider the part of the cycle containing vertices $j_i$, $j_{i+1}$ and sides $x_{i-1}$, $x_i$, $x_{i+1}$. From the previous lemma, and from the procedure used to construct $\Gamma_{\varphi\psi}$, it follows that vertex $x_{i-1} \in \tilde{X}$ on the periodical section of the trajectory with period $1^i_\psi$, is transferred to the complex elements of cycle $X^{*i}_\psi$. In particular, after each $1^{i+1}_\psi$ steps, it is subordinated to vertex $x_i$. For the same reason, vertex $x_i$ is subordinated to vertex $x_{i+1}$ after $1^{i+1}_\psi$ steps. As $1^i_\psi$ and $1^{i+1}_\psi$ are not multiple, a subordination chain $\eta_t(x_1) = x_2, \ldots \ldots \eta_t(x_{k-1}) = x_k$ is formed aftr $d_i$ steps, where d is the least common multiple of numbers $1^i_\psi$ and $1^{i+1}_\psi$. Using the same procedure on the whole chain, the result will be that of the subordination chain $\eta_t(x_k) = x_1$; $\eta_t(x_1) = x_2, \ldots,$ $\eta_t(x_{k-1}) = x_k$, that is, the complex cycle of graph $\Gamma\eta(t)$, including vertices $x_1, x_2, \ldots x_k$ is formed after D steps, where D is the least common multiple of numbers $1^i_\psi \in L$.

We assume that $j_n$ and $j_m$ are the nearest two vertices of the cycle, with multiple lengths $1^{jn}_\psi$ and $1^{jm}_\psi$. In addition, the lengths of lines of graph $\Gamma_\psi$, making the projection $\pi(x_{n-1})$ and $\pi(x_{m-1})$ correspond to elements $x_n$ and $x_m$, are:

$$1_{x_{n-1}} = p1^{jn} + s^{xn-1}_{xn}, \quad 1_{x_{m-1}} = q1^{jm}_\psi + s^{xm-1}_{x_m}$$

and because of the equality of indices

$$s^{x\,n-1}_{x_n} = s^{xm-1}_{x_m}.$$

We will assume $1_{x_{n-1}} > 1_{x_{m-1}}$.

In this case, in graph $\Gamma\eta(t)$, subordination $\eta_t(x_{m-1}) = x_m$, $t = 1_{x_{m-1}}$ is performed first, and then, after $\Delta 1 = 1_{x_{n-1}} - 1_{x_{m-1}}$ steps, subordination $\eta_t(x_{n-1}) = x_n$, $t = 1_{x_{n-1}}$.

As they are multiples, $1^{jm}_\psi = \mu\, 1^{jm}$ or $1^{jm}_\psi = \mu\, 1^{jn}$, where $\mu$ is an integer. We assume at first that $1^{jn}_\psi = 1^{jm}_\psi$; then $\Delta 1 = (p\mu - q)1^{jm}_\psi$, that is, the subordination $\eta_t(x_{n-1}) = x_n$ and the $(p\mu - q + 1)$-th subordination $\eta_t(x_{m-1}) = x$ are carried out simultaneously, and furthermore, will be carried out after $1^{jn}_\psi$ steps.

If $l^{j_m}{}_{\psi} = \mu l^{j_n}{}_{\psi}$, then $\Delta l = (p - g\mu) l^{j_n}_{\psi}$ and subordination $\eta_t(x_{m-1}) = x_m$ and the $((g+1)\mu - p + 1)$-th subordination $\eta_t(x_{m-1}) = x_m$ are carried out simultaneously and, furthermore, they will be carried out after $l^{j_m}{}_{\psi}$ steps. Evidently, the same situation is true for $l_{x_{m-1}} \geq l_{x_n}$. Therefore, subordination chain $\eta_t(x_{n-1}) = x_n$, $\eta_t(x_n) = x_{n+1}, \ldots, \eta_t(x_{m-1}) = x_m$ in graph $\tilde{\Gamma}_{\eta}(t)$ is carried out every D' steps, where D' is the least common multiple of sequence $l^{j_n}{}_{\psi}, l^{j_{n+1}}_{\psi}, \ldots, l^{j_m}_{\psi}$. When all the vertices with multiple lengths of the cycle in question have the property of equality of indices, then the whole subordination chain $\eta_t(x_1) = x_2$, $\eta_t(x_2) = x_3, \ldots$ $\ldots, \eta_t(x_k) = x_1$, that is, cycle $x_1, x_2, \ldots, x_k$ in graph $\tilde{\Gamma}_{\eta}(t)$ closes after D steps, where D is the least common multiple in L. Note that, if a $p(x_s)$ chain $x_s \in \tilde{X}$ contains more than one element, this chain will completely enter the corresponding cycle of graph $\tilde{\Gamma}_{\eta}(t)$.

The theorem is therefore sufficient.

Proof of necessity.

Element x does not enter the set of sides of graph $\Gamma_{\varphi\psi}$ in one of the following cases:

1) Element $x \in X^*$ does not have a projection,

2) element $x \in X^*_{\psi}$ has a projection but the corresponding side in graph $\Gamma'_{\varphi\psi}$ does not enter an oriented cycle satisfying the equality conditions of indices,

3) the element does not enter any $p(x^*)$ chain of an element $x^* \in X^*_{\psi}$, which has a side in $\Gamma_{\varphi\psi}$

In case 1, x is constantly found among the non-complex elements of the trivial component and therefore, it never appears in set $X^*\eta(t)$.

In case 2, side x of graph $\Gamma_{\varphi\psi}$ either enters an oriented non-closed chain of vertices $j_1, j_2, \ldots j_k$ linked by sides $x_1, x_2, \ldots x, \ldots, x_k$ or, even if x enters a cycle, this cycle contains at least a pair of vertices, so that the equality condition of indices is not satisfied.

From the proof of sufficiency, it follows that neither in the first, nor in the second case, the chain of sides containing x is closed. Therefore, x cannot enter any complex cycle of graph $\Gamma_{\eta}(t)$ (together with its chain).

In case 3, two variants are possible:

a. $x \in X_{\psi}$ then x is a hanging vertex in $\Gamma_{\eta}(t)$

b. $x \in X_{\varphi} \setminus X_{\psi}$

As $p(x^*)$, $x^* \in X^*_{\psi}$ does not belong to any chain, this means that x is the element of a trivial component and further, the chain, connecting x with its complex cycle, does not enter $X'_{\psi}$. According to the lemma, the whole chain belongs to the trivial component at any step, and therefore, $x \in E_T(x^* \eta(t))$.

This proves the necessity of theorem 2.5.

Graphs $\Gamma'_{\varphi\psi}$ and $\Gamma_{\varphi\psi}$ can be constructed because of example 2.4. p(x) chains corresponding to the complex vertices of graph $\Gamma_{\psi}$ and the sets to which their projections belong are shown in Table 2.5.

Fig. 2.9 shows graphs $\Gamma'_{\varphi\psi}$ (Fig. 2.9a) and $\Gamma_{\varphi\psi}$ (Fig. 2.9b) which have been constructed according to table 2.5. According to theorem 2.5, it follows, from Fig. 2.9b, that $E_T(X^*_{\eta}(t)) = \{m, j, k, w, i, g, z\}$ and that together with complex elements $\{m, k, w, z\}$ the envelope also contains the elements of their chains. This deduction is confirmed by the trajectory in Fig. 2.8c.

| Component | Complex Element $x \in X^{*j}_{\psi}$ | Chain p (x) | Projections belonging to sets $\pi(x)$ |
|---|---|---|---|
| $\Gamma_{\psi_1}$ | ж<br>К | p(ж) = {ж, j}<br>p(К) = {К} | π(ж) ∈ $X^1_\psi$<br>π(ж) ∈ $X^4_\psi$ |
| $\Gamma_{\psi_2}$ | з<br>ц | p(з) = {з} | π(з) ∈ $X^4_\psi$ |
| $\Gamma_{\psi_3}$ | ф<br>Т<br>э | p(Т) = {Т η}<br>p(э) = {э} | π(Т) ∈ $X^4$<br>π(э) ∈ $X^3_\psi$ |
| $\Gamma_{\psi_4}$ | щ | p(щ) = {щ i g} | π(щ) ∈ $X^1_\psi$ |

Table 2.5

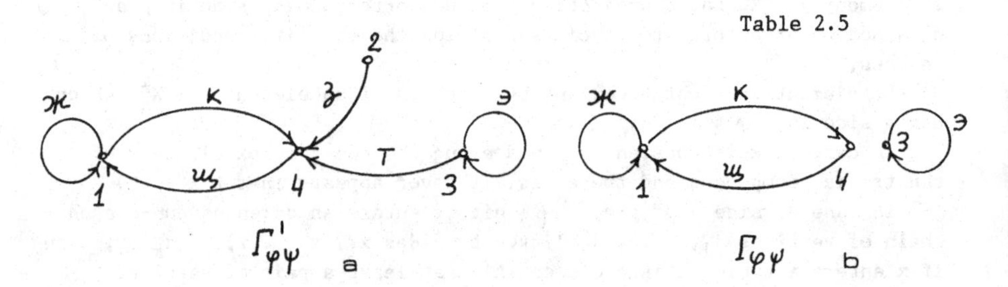


Fig. 2.9

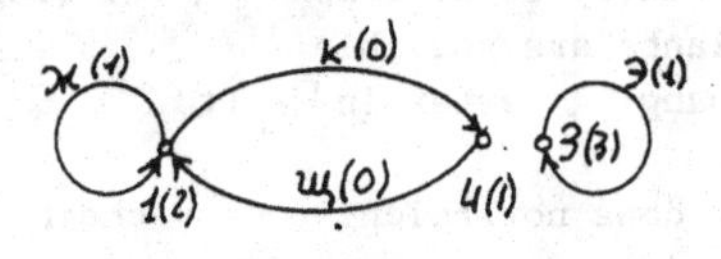


Fig. 2.10

From theorem 2.5, we have the following corollary.

Corollary 2.5 - 1

Any element of the collective envelope $E_T(X^*_\eta(t))$ belongs to the set $\{p(x) \setminus x \in X^*_\psi$ of the elements of the chain of the complex subset.

Obviously, vertices on graph $\Gamma_{\varphi\psi}$ do not always correspond to an active component of the graph, for instance, in Fig. 2.9 there is no vertex 2 corresponding to the active component $\Gamma_{\psi_2}$ of graph $\Gamma_\psi$.

Corollary 2.5 - 2

Period T* of complex subset $X^*_\eta(t)$ is equal to the least common multiple of the vertices of graph $\Gamma_{\varphi\psi}$.

Evidently, any substitution operator $\Gamma_\psi$ produces a partition of the set of CH-graphs into classes $G_1, G_2, \ldots$, such that one and only one complex envelope of the sequence of graph 2.6 corresponds to any initial graph $\Gamma_\varphi \in G_i$

## 2.5 Constant and Variable components of a complex envelope - Structure and Stability

In a given complex envelope $E_T(X^*_\eta(t)$ of a periodical solution of equation 2.5, various elements $x \in E_T(X^*_\eta(t))$ may appear in a complex set $X^*_\eta(t)$ of the periodical solution during different numbers of steps.

Set $E_T(X^*_\eta(t)) \subseteq (E_T(X^*_\eta(t))$ where each element belongs to the complex set $X^*_\eta(t)$ for any $t \in [0, T]$, is called constant component of the complex envelope $E_T(X^*_\eta(t))$ and set $\tilde{E}_T(X^*_\eta(t)) = E_T(X^*_\eta(t)) \setminus \bar{E}(X^*_\eta(t))$ variable component.

To describe the complex envelope, we considered an auxiliary digraph $\Gamma_{\varphi\psi}$. A connected component of graph $\Gamma_{\varphi\psi}$ is said to be constant if, for any vertex i belonging to this component, the following conditions are valid:

1) the number of sides convergent to vertex i is equal to its index $l^i_\psi$, that is, it is equal to the length of the complex cycle of graph $\Gamma_\psi$ which corresponds to this vertex;

2) For all elements having sides convergent to vertex i, the indices of the projections, with respect to some complex elements of the i-th connected component to graph $\Gamma_\psi$, are different.

<u>Theorem 2.6</u>

The constant component of complex envelope $\hat{E}_T(X^*\ \ (t))$ consists only of those elements which have their sides in the constant components of graph $\Gamma_{\varphi\psi}$ and of the elements of their chains.

<u>Proof</u>

We assume that $x_1, x_2, \ldots, x_k$ are the sides of the constant component of graph $\Gamma_{\varphi\psi}$ convergent to vertex i with length $l^i_\psi = k$ and $x'_1, x'_2, \ldots x'_k$ are the sides coming from this same vertex. By virtue of the lemma, the inequality of the indices of all convergent sides, implies that the elements which correspond to them can never enter a complex, and, therefore, they are, at any moment, all subordinated to different elements $x'_1, x'_2, \ldots, x'_k$ (even by means of their $p(x_s)$ chain, $s = 1, 2, \ldots, k$).

If $\tilde{X}$ is the set of all the sides of the connected component in question, and if all their p(x) chains, $x \in \tilde{X}$ consist only of the same elements x, then it follows that $\varphi(\tilde{X}) = \tilde{X}$, that is, all elements $x \in \hat{X}$ always belong to the complex set of graph $\tilde{\Gamma}_\eta(t)$. A similar discussion is also valid for the elements of p(x) chain with length longer than 1.

Sufficiency is thus proved.

We will prove that in a connected component of graph $\Gamma_{\varphi\psi}$ which is not constant, there is always a vertex of one of the two following types: type 1, with all entering sides and type 2, with all exiting sides, which are periodically excluded from the complex set of graph $\tilde{\Gamma}_\eta(t)$.

In fact, if we are not dealing with the constant component, this will contain at least one vertex for which, either a) the number of exiting sides is smaller than the index of this vertex, or b) the number of sides entering it is smaller than its index, or c) the number of entering sides is equal to the index of the vertex, but there is at least one pair of sides with equal indices between them, or d) the number of entering indices is larger than the index of the vertex.

In case a), the element without a side exiting from the vertex does not belong to the complex envelope, but each element corresponding to entering sides is subordinated to this element for no less than one step of the period, i. e., in that moment, they leave the complex set and therefore belong to the variable component.

In case b), no element which corresponds to entering sides for at least one step in the period, is subordinated to any of the elements correspondent to the complex set.

In case c), according to the lemma, a pair of elements corresponding to the exiting sides with equal indices always belongs to the same complex and therefore, as in case b), all the elements corresponding to the exiting sides periodically leave the complex set.

Finally, case d) means that this component has another vertex so that case b) is applied. Let x now be an arbitrary side belonging to a non-constant component; as a result of the connectivity property, a chain can always be found which includes a sequence of vertices with numbers $n_1, n_2, \ldots\ldots, n_k$ connected with a sequence of sides $x_1, x_2, \ldots, x_{k-1}$ without repetitions (here side $x_1$ exits from vertex $n_1$ and enters vertex $n_{i+1}$) and in addition, either $n_1$ is of type 1 and then side $x_1$ exits from $n_k$, or vertex $n_k$ is of type 2, and then x enters $n_1$. It is easy to see that subordination chain $\eta_t(x_1) = x_2$, $\eta_t(x_2) = x_3, \ldots, \eta_t(x_{k-1}) = x$, in case $n_1$, is of type 1), or chain $\eta_t(x) = x_1$, $\eta_t(x_1) = x_2, \ldots, \eta_t(x_{k-1})$, when $n_k$ is of type 2, exits from the complex subset of $\Gamma_\eta(t)$ in the periodical section of the trajectory, after every D steps, where D is the least common multiple of quantities $1^{n1}_\psi$, $1^{n2}_\psi, \ldots\ldots, 1^{nh}_\psi$ which correspond to vertices $n_1, n_2, \ldots, n_k$.

Thus necessity is proved.

Theorem 2.6 defines the set of all the vertices of graphs $\Gamma_\varphi$ and $\Gamma_\psi$ which constantly remain as elements of the complex envelope of graph $\Gamma_\eta(t)$, that is, they enter some of its complex cycles.

Example 2.5 In Fig. 2.10, we will re-consider graph $\Gamma_{\varphi\psi}$ constructed for example 2.4. For vertices, in parentheses we give their indices, and for sides, the indices of the corresponding projections (with respect to "$\mu$" for vertex 1, to "$\nu$" for vertex 4 and to "$\phi$" for vertex 3).

Evidently, the connected component with vertices 1 and 4 is constant, whereas the component with isolated vertex 3 is not. Therefore, in a periodical solution, vertices $\varkappa$, $\kappa$, $\mu$ enter complex set $X^*_\eta(t)$ together with the elements of their chains at any step, while vertex "$\ni$" belongs to it only at some steps (see the trajectory in Fig. 2.8b and Table 2.4).

The evaluation of the stability of element $x \in E_T(X^*_\eta(t))$ can be given by the stability coefficient $k_x$ which is determined by the frequency of the element's appearance in a complex set:

$$k_x = \frac{m\ x}{T}$$

where m is the number of times in period T that the element appears in a complex set: $x \in X^*_\eta(t)$.

Not only can the constant of the complex set vary along the trajectory of the periodical solution, but also its structure: some complex cycles may disintegrate, others may be formed. Different cycles of the structure may appear with different frequency and the stability of arbitrary cycles is evaluated from the stability coefficient of cycle

$$k_{c_i} = \frac{\tau_{ci}}{T}$$

where $\tau_{c_i}$ is the number of times in a period T that cycle $c_i$ is in the complex cycle.

The stability of any element in a complex envelope cannot be less than the stability of the cycle to which this element belongs. When element x enters several cycles, its stability coefficient is equal to the sum of the stability coefficient of these cycles.

Set $\Gamma''_{\varphi\psi}$ denoting the whole of the oriented cycles contained in graph $\Gamma_{\varphi\psi}$ has cycles of two types: a) simple cycles, not containing the same vertex more than once; b) cycles with multiple vertices.

In Fig. 2.10, we show an example of such a type: 4 $\mu \rightarrow$ 1 $m \rightarrow$ 1k $\rightarrow$ 4 (numbers refer to vertices and letters to sides). Once having checked, for every cycle, the equality condition of indices, and deleted those cycles not satisfying this condition, the remaining cycles can be transformed as follows: each side with index x is substituted by the p(x)chain and the vertices by oriented sides which connect these chains. The whole of isolated cycles thus obtained is denoted by $\Gamma^*_{\varphi\psi}$.

__Theorem 2.7__

a) The structure of the complex envelope of the periodical solution of equation 2.5 coincides with $\Gamma^*_{\varphi\psi}$;

b) stability coefficient $k_{c_i}$ of complex cycle $c_i \in \Gamma^*_{\varphi\psi}$ is determined by the expression $k_{c_i} = 1/D_i$, where $D_i$ is the least common multiple of the indices of vertices of the correspondent cycle in $\Gamma^*_\psi$.

c) Stability coefficient $k_{x_j}$ of element $x_j \in E_T(X^*_\eta(t))$ is expressed by

$$k_{x_j} = \sum_i k_{c_i}$$

where the condition is extended to all cycles $c_i$ where $x_j$ enters.

Proof

According to the lemma, the structure of graph $\Gamma_{\varphi\psi}$ determines the structure of the complex envelope of graph $\tilde{\Gamma}_{\eta}(t)$ in the following manner: when side $x_1$ enters vertex i of graph $\Gamma_{\varphi\psi}$ and side $x_2$ exits from the same vertex, then, in graph $\Gamma_{\eta}(t)$, only the subordination of element $x_1$ to element $x_2$, and not the contrary, is possible. Therefore, the structure of the complex envelope is included in the set of all the possible oriented cycles of graph $\Gamma_{\varphi\psi}$. Let $n_1, n_2, \ldots, n_k$ be a sequence of vertices of some oriented cycle of graph $\Gamma_{\varphi\psi}$, $x_1, x_2, \ldots, x_{k+1}$ a sequence of sides belonging to this cycle, so that side $x_1$ enters vertex $n_1$ and side $x_{i+1}$ exits from the same vertex. Now, both when the chain does not contain vertices with multiple lengths, and when such vertices exist but they satisfy the equality condition of indices, the formation of subordination chain $\eta_t(x_1) = x_2$, $\eta_t(x_2) = x_3, \ldots, \eta_t(x_k) = x_{k+1}$ in graph $\tilde{\Gamma}_{\eta}(t)$ is obtained after D steps, where D is the least common multiple of the lengths of vertices $1^{n1}_{\psi}, 1^{n2}_{\psi}, \ldots, 1^{nk}_{\psi}$. During the period, this formation is obtained $\frac{T}{D}$ times. If the equality condition of indices were not respected, then such a chain would not be completely formed; hence follow the statements a) and b) of the theorem.

Point c) is a result of the fact that the elements of the complex envelope may appear in complex set $X^*_{\eta}(t)$ of graph $\tilde{\Gamma}^*_{\eta\psi}$ in only one of cycles $\Gamma^*_{\varphi\psi}$, and that two different cycles, containing the same element, are formed at different moments.

As each of the components contains exactly one cycle, theorem 2.7 - makes it possible to construct all the possible complex cycles on the periodical trajectory of graph $\tilde{\Gamma}_{\eta}(t)$ - determines at the same time, both the number of all the different complex components, appearing on the trajectory at different moments, and the type of these components. In addition, this theorem allows the calculation of the frequency of different cycles which are formed, and how frequently these elements belong to the cycle. From the construction procedure of graph $\Gamma^*_{\varphi\psi}$ and theorem 2.7, it follows that complex cycles in the structure of envelope $E_T(X^*_{\eta}(t))$ certainly correspond to all the pure cycles of

Example 2.6 Consider, now, graph $\Gamma_{\psi}$ in Fig. 2.11a and initial graph $\Gamma_{\varphi}$ in Fig. 2.11b. In Table 2.6, we give the p(x) chains for all complex elements $x \in X^*_{\psi}$. Letters i, j $\notin X_{\psi}$, for which component $\Gamma_{\varphi 2}$ of initial graph is trivial. That all the components of substitution graph $\Gamma_{\psi}$ are active, is demonstrated by the fact that all projections of complex elements $x \in X^*_{\psi}$ belong to components $\Gamma_{\psi}$ (see the last column in Tab. 2.6). Thus the period of the solution is T = T max = 6.

In Fig. 2.12a, we show graph $\Gamma'_{\varphi\psi}$ constructed according to Table 2.6, and in Fig. 2.12b, graph $\Gamma_{\varphi\psi}$, obtained from $\Gamma'_{\varphi\psi}$ through the procedure described in paragraph 2.4. From Fig. 2.12b and Table 2.6, it follows that $E_T(X^*_{\eta}(t)) = \{T\ d\ \gamma\ u\ \phi\ \tau\}$. Graph $\Gamma_{\varphi\psi}$ contains only one connected component, to which vertex 2 of length $1_2 = 2$ belongs, and on which only a side is incident.

Therefore, this component is not constant, and the complex envelope does not have a constant component.

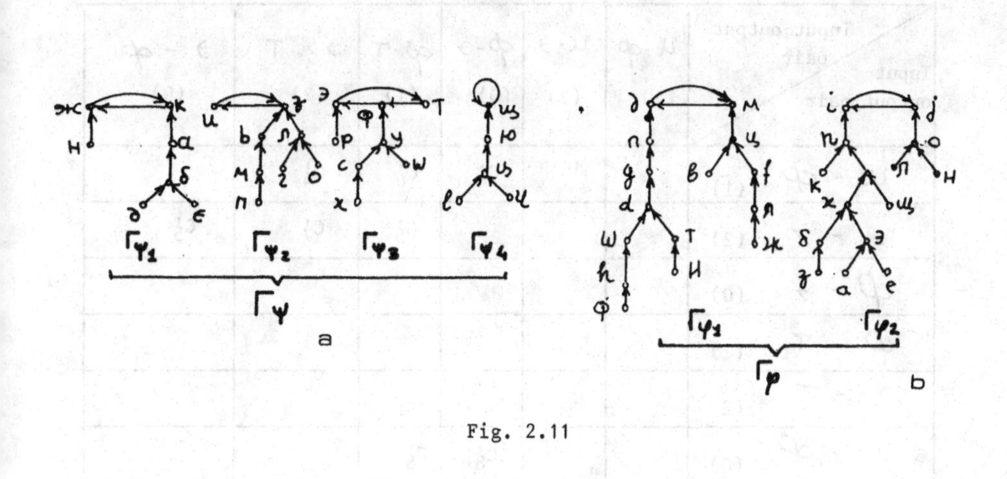


Fig. 2.11

| Complex Element | Chain | Belonging of projection $\pi(x)$ to the components of $\Gamma_\psi$ |
|---|---|---|
| $\text{ж} \in X^J_\psi$ | $p(\text{ж}) = \{\text{ж a f}\}$ | $\pi(\text{ж}) \in X$ |
| $\text{з} \in X^2$ | $p(\text{з}) = \{\text{з}\}$ | $\pi(\text{з}) \in X^1$ |
| $\text{и} \in X^2$ | $p(\text{и}) = \{\text{и}\}$ | $\pi(\text{и}) \in X^3$ |
| $\text{к} \in X^1$ | - | - |
| $\text{т} \in X^3$ | $p(\text{т}) = \{\text{т d ч}\}$ | $\pi(\text{т}) \in X^2$ |
| $\text{ф} \in X^3$ | $p(\text{ф}) = \{\text{ф h}\}$ | $\pi(\text{ф}) \in X^3$ |
| $\text{щ} \in X^4$ | $p(\text{щ}) = \{\text{щ}\}$ | $\pi(\text{щ}) \in X^2$ |
| $\text{э} \in X^3$ | $p(\text{э}) = \{\text{э}\}$ | $\pi(\text{з}) \in X^3$ |

Table 2.6

| Input-output pair / Input-output pair | И-Ф (1) | И-Э (2) | Ф-Э (0) | Ф-Т (1) | Э-Т (2) | Э-Ф (0) |
|---|---|---|---|---|---|---|
| И - Ф (1) | | | $C'_4$ | $C'_6$ | $C'_4$ | |
| И - Э (2) | | | | $C'_5$ | $C'_7$ | $C'_5$ |
| Ф - Э (0) | | | | | | |
| Ф - Т (1) | | | | | | |
| Э - Т (2) | | | | | | |
| Э - Ф (0) | | | $C'_8$ | $C'_5$ | | |

Table 2.7

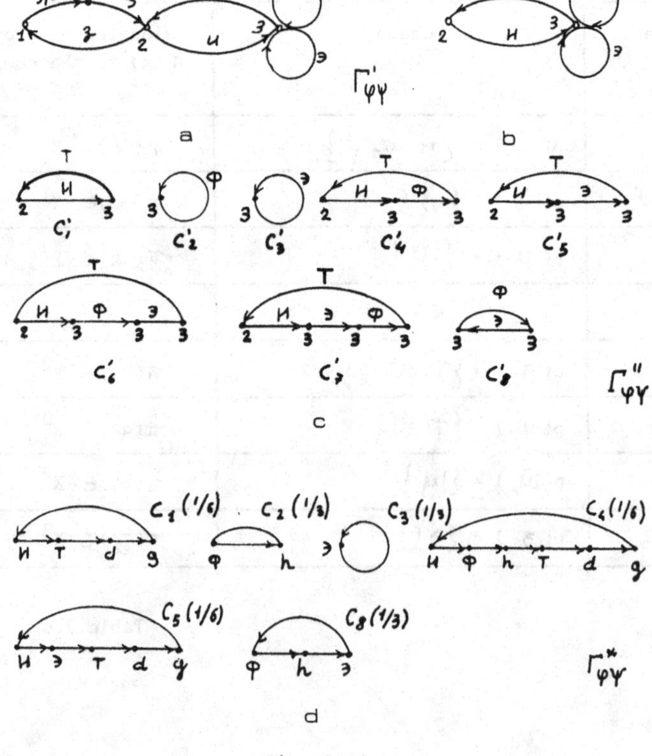


Fig. 2.12

In addition, Fig. 2.12c shows the set of cycles $\Gamma''_{\varphi\psi}$. Cycles $C'_1$, $C'_2$ and $C'_3$ are simple, the remaining cycles contain multiple vertex 3. For the analysis of these cycles we go back to Table 2.7.

On this Table, a line and a column (in parentheses we give the index of the entering side with respect to the exiting one) at whose intersection we find the cycle with the combination of such a pair, correspond to every pair of "entering-exiting sides" of vertex 3, found in cycles $C'_4$ — $C'_8$. The Table shows that cycles $C'_6$ and $C'_7$ do not satisfy the equality condition of indices and therefore, they must not be considered for the construction of graph $\Gamma^*_{\varphi\psi}$. In Fig. 2.12d, we show graph $\Gamma^*_{\varphi\psi}$, which represents the structure of the envelope obtained from Fig. 2.12c and Table 2.7.

Every complex cycle has its own stability coefficient next to it in parentheses, and Table 2.8 shows the stability coefficients of the complex elements of the correspondent cycles.

| Complex Element | u | T | ф | э | a | y | h |
|---|---|---|---|---|---|---|---|
| Stability Coefficient | 1/2 | 1/2 | 5/6 | 5/6 | 1/2 | 1/2 | 1/2 |

Table 2.8

Three corollaries of theorem 2.7 supply us with some evaluations of the structure of the complex envelope without a detailed analysis being necessary.

<u>Corollary 2.7 - 1</u>

The highest number of the connected components of periodical solution $\tilde{\Gamma}_\eta(t)$, for a given substitution graph $\Gamma_\psi$ is equal to the number of its complex vertices.

<u>Corollary 2.7 - 2</u>

Only those complex vertices x of graph $\Gamma_\psi$, a) which have projections on their own components; b) whose p(x) chains consist exclusively of those same elements, can be roots of components T of the periodical solution.

<u>Corollary 2.7 - 3</u>

Cycles of the complex envelope having stability coefficient equal to 1 consist only of root x of components T of graph $\Gamma_\psi$ and of their elements of p(x) chains.

Theorem 2.6 makes it possible to determine the set of elements $x \in X$ which are constantly complex along the trajectory of periodical solution $\tilde{\Gamma}$ (t), even if they may belong to different cycles.

Corollary 2.7 - 3 determines the set of cycles which are constantly complex along the trajectory.

In this chapter the properties of the equations with CH-graphs have been discussed. We have shown that these equations always have a stationary or periodical solution, which can be obtained after a finite number of steps. Some statements have been proved, which allow the deduction of the properties of the solutions from the properties of the initial graph and from the graph of the substitution operator. It is now clear that some properties are completely determined by the substitution graph, independently of the initial graph; such are the set of possible periods and the highest number of connected components in the periodical solution.

The remaining properties do not depend exclusively on the substitution graph but also on the initial graph; such are the period of the solution, the complex envelope, its constant component and its structure, i. e., all complex cycles which sometimes appear during the period, and so on.

Naturally, it is obvious that the class of substitution operators does not exhaust the set of all possible types of graph operators.

## 3. DESCRIPTION OF DYNAMIC STRUCTURES THROUGH WEB GRAMMARS

### 3.1 Web Grammars

In this chapter we will propose another language for the description of problems of dynamics of graphs, that is, the language of Web grammars. A Web grammar contains a finite set of initial graphs and a finite set of rules for admissible local graph transformations.

The grammar is transformational, if the set of initial graphs is not specified; due to its own substitution rules, it transforms the assigned graph when such transformations are possible.

This method for the description of classes of graphs and of their changes in discrete times, gives rise to and solves some new problems.

Whereas the operations on graphs introduced in chapter I allow the univocal extension of a trajectory, it is possible here to obtain the branching of a trajectory, since many grammar rules can simultaneously be applied to a graph.

In this chapter we will present the solution to the problem of finding a class of equilibrium graphs or those which converge to equilibrium.

The language of Web grammars seems to be convenient to describe the changes in real objects, if the change is considered as the independent change of its parts; (when the object is so large that the description of its global re-organization becomes impossible).

Unlike Chomsky's grammars, grammars have been introduced into literature fairly recently, mainly in connection with the description of classes of images.

We will start with a brief list of the basic characteristics of Web grammars, which will be then used to obtain some results specifically connected with the dynamics of graphs.

As it is well-known, formal grammars are used to create (and transform) chains of symbols (words) from a finite group of symbols (alphabet). The set of all the words formed from an alphabet V is denoted by $V^*$.

Definition 3.1

A formal grammar is defined by the quadruple $G = \langle V. T. P. S \rangle$, where V and T are finite alphabets ($T \subset V$, T is a terminal alphabet and $V \setminus T$ a non-terminal alphabet), S is the initial symbol ($S \in V \setminus T$) and P is a finite collection of substitution rules of type $\varphi \to \Psi$ (where $\varphi, \Psi \in V^*$ are words); the length of word $\varphi$, is larger than zero and $\varphi$ contains at least one symbol of $V \setminus T$. We say that word $\omega'$ derives directly from $\omega$ in grammar G, if:

1) $\omega = X \varphi \xi$, $\omega' = X \Psi \xi$, where $X, \varphi, \Psi, \xi \in V^*$;
2) rule $\varphi \to \Psi$ belongs to P.

Direct derivability relation is denoted by $\underset{G}{\Rightarrow}$. Derivability relation $\overset{*}{\underset{G}{\Rightarrow}}$ is the reflective and transitive closure of direct derivability relation $\underset{G}{\Rightarrow}$. The language generated by grammar G is

$$L(G) = \{ \omega / \omega \in T^*, \quad S \overset{*}{\underset{G}{\Rightarrow}} \omega . \}$$

In other words, the set of all the words derivable from S, using the rules of the grammar, is the grammar generated using grammar G. Usually, constraints concerning the choice of substitution rules appear in formal grammars; they are divided into classes according to the substitution rules contained, and the imposed constraints (see ref. [3], for instance).

From the point of view of graph grammars, we can "graphically interpret" a chain if any word is thought as being an oriented graph with indices on the vertices. From this point of view, the application of rule $\varphi \to \Psi$ to the graph representing word $\omega$ is reduced to the implementation of the following algorithm: "Separate the edges connecting subchain $\varphi$ with the remaining vertices of the graph which represents $\omega$; connect the vertex of chain $\Psi$, according to the first letter of word $\Psi$, with an oriented edge. The vertex of this was connected with the vertex of graph $\varphi$, which corresponds to the first letter of word . then connect then vertex corresponding to the last letter of $\varphi$, with an oriented edge, at whose vertex the vertex corresponding to the last letter of $\Psi$ was connected".

We will denote by $\{\Gamma\}$ the set of vertices of graph $\Gamma$. The term "subgraph" of graph $\Gamma$ is used to denote a graph $\Gamma'$, such that $\{\Gamma'\} \subseteq \{\Gamma\}$ The terms "proximity", "distance" and "neighbourhood of a ray" are used in their normal sense; for instance, the distance between two vertices of a graph is the length of the shortest path connecting them.

In graph grammars, as in usual grammars, we use a finite group of rules and two finite non-intersecting alphabets (both non-terminal and terminal). Their union represents the general alphabet of the grammar.

The two members (right and left) of a rule in such a grammar are graphs of arbitrary form. Each rule has its own algorithm which makes it possible to add the right member to the graph instead of the left. As in usual grammars, there must be symbols taken from the non-terminal alphabet among the indices of the right member of the rule; indices can be assigned both to the vertices and to the sides of a graph.

In such a grammar, a step consists in the substitution of a local part of a graph with another graph with the help of one of the grammar rules; the number of steps in a derivation represents its length. For our purposes, it is convenient to represent the graph such that close vertices (in the sense of the distance in a graph) are represented close to each other; then each step will appear as the substitution of a small part of a graph with another diagram.

A grammar for G graphs generates a language L(G) which consists of all the graphs derivable from the initial graph through the rules of grammar G.

An example of effective application of graph grammars can be the well-known proof of Kleene's theorem, from the fact that for any regular expression, there is a finite automaton which recognizes the language defined by this expression. This proof is constructed by using a grammar which generates the block - diagram of the finite automaton for an arbitrary regular expression. The use of the language of grammars for graphs to prove Kleene's theorem is important, because it allows the immediate construction of the block - diagram of the automaton during the proof.

Contrary to usual grammars, specific problems arise in graph grammar which are connected with the formation of all the graphs belonging to some specific classes; e. g., the class of all non-separable graphs, of all acyclic oriented graphs, of all trees, etc.

When we tackle such problems, the indices of the vertices have an arbitrary role (the terminal alphabet often consists of a single symbol).

The models for graph grammar are divided into two classes, each of them connected with one of the interpretations of the term "graph". A graph can be treated both as a set of sides, with common nodes, and as a set of vertices connected by sides.

In common graph theory, both interpretations have the same formal results, the difference arising only for more complex objects, for example, hypergraphs.

In graph grammar, the method of substituting a graph $\Gamma'$ in graph $\Gamma$ with a given graph $\Gamma''$ depends on the interpretation of the term "graph". Such a substitution presupposes the removal of subgraph $\Gamma'$. The cut dividing $\Gamma'$ from $\Gamma$ can be made either with respect to the vertices or the sides connecting $\Gamma'$ and $\Gamma$. As a result, two main models of graph grammar can be examined: side grammars and vertex grammars or Web grammars.

Since problems of graph dynamics are connected with vertex grammars, we will deal with this in the next paragraph, - (for the grammar of the sides see ref. [4]).

<u>Web Grammars</u>

In a Web grammar G, when a subgraph $\Gamma'$ in graph $\Gamma$ is substituted by a graph $\Gamma''$, all the vertices belonging to $\Gamma'$ and all the sides (or edges, if we are dealing with an oriented graph) having at least one extreme in incidence on vertex $\Gamma'$ are removed from $\Gamma$. Graph $\Gamma''$ is dipped into $\Gamma$ with the aid of a special algorithm E, which forms the sides linking the vertices of $\Gamma''$ to the vertices of $(\{\Gamma\}\setminus\{\Gamma'\})$.

Since algorithm E links the vertices of $\Gamma''$ to the vertices of the neighbourhoods[3] of $\Gamma'$ only, changes in the graph are of a local nature.

<u>Definition 3.2</u>

A quadruple G = $< V, \Sigma, P \;\; I >$ is called a Web grammar, where V is the union of the terminal and non-terminal alphabet, $\Sigma$ is the terminal alphabet, $\Sigma \subseteq V$, P is the finite set of substitution rules and I the finite set of initial graphs with indices. Every substitution rule has the form $\alpha$ , C $\to \beta$ , E where $\alpha$ and $\beta$ are graphs whose vertices are marked by symbols taken from V, and among the indices on the vertices of $\alpha$ there are symbols taken from $V \setminus \Sigma$ ; C is a logical condition of applicability; E is an immersion algorithm. We say that graph $\Gamma'$ is derived from $\Gamma$ ($\Gamma \Rightarrow \Gamma'$) with the aid of rule ($\alpha$ , C, $\beta$ , E) is a subgraph isomorphic with $\alpha$ can be found in $\Gamma$ . If C is true and if graph $\Gamma'$ is obtained from $\Gamma$ as a result of the subordination of $\alpha$ with $\beta$ in the construction of a side between the vertices of $\beta$ and the vertices of $O_\alpha \setminus \{\alpha\}$ according to the instructions of E.

The condition of applicability C is linked to the notion of "negative context". When, in normal grammars, it is required that the substitution of a chain $\varphi$ with a chain $\Psi$ is to be performed only when a chain $\omega$ is not found on the left of $\varphi$ and a chain $X$ not on the right, it is sufficient to introduce a group of rules $\{\xi\varphi\eta \to \xi\Psi\eta\}$ where $|\xi| = |\omega|$, $|\eta| = |X|$, $\xi$ and $\eta$ are all the possible chains, besides $\omega$ and $X$ . In graph grammars it is not possible to do the same thing, as the number of possible neighbourhoods of a graph is not limited, and consequently we should introduce an infinite number of rules in order to specify the substitution of $\alpha \to \beta$ , E on condition that a subgraph is not placed in the neighbourhood of $\alpha$ .

However, the following theorem is valid.

Theorem 3.1

For any Web grammar G, an equivalent grammar C with a condition of applicability C true for any rule can be constructed.

Mantovani has formulated and proved this theorem [5].

The condition of applicability of a rule is included within the definition of a Web grammar, in order to point out the substantial difference between it and the ordinary ones. However, due to theorem 3.1, we will use only the definition of a Web grammar, where the condition of applicability C is not included in the substitution rules.

Example 3.1 Let us consider grammar $G_1$ = < V, $\Sigma$ , P, I > where V = {A, a}, $\Sigma$ = {a} , I = {. a }, I is the initial graph shown in Fig. 3.1a and P is the system of the two rules represented in Fig. 3.1b.

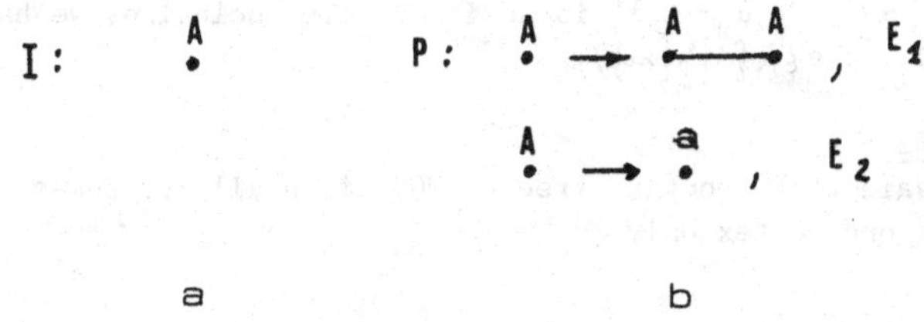

I: A•        P: A• → A•——•A , $E_1$

A• → a• , $E_2$

a        b

Fig. 3.1

Algorithm $E_1$ links the vertices on the right side with all the vertices of $0_A$, where A is the left side of the first rule of Fig. 3.1b. Algorithm $E_2$ links vertex a with all the vertices of $0_A$, where A is the left side of the second rule of Fig. 3.1b. It is clear that $L(G_1)$ consists only of all the complete graphs. It is also evident that the length of a generation of a complete graph with n vertices is equal to 2n.

Fig. 3.2 shows a complete generation of a graph with four vertices.

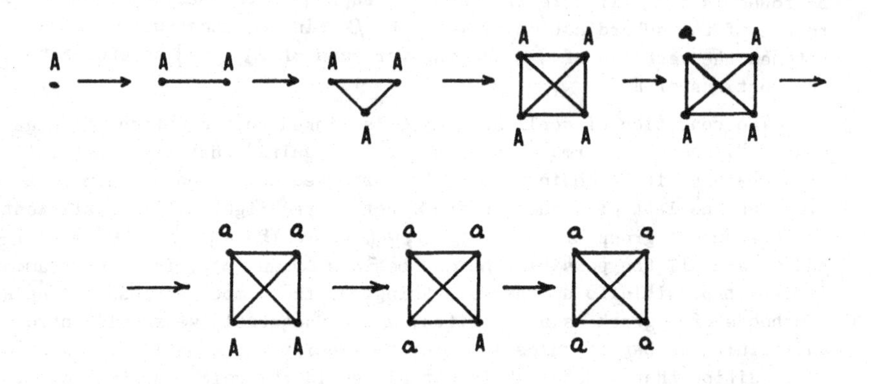

Fig. 3.2

The algorithm for the substitution rule $\alpha \rightarrow \beta$ , E can be given with the aid of correspondence f: $\{\alpha\} \rightarrow \{\beta\}$ . Such an algorithm connects a vertex $u \in \{\beta\}$ with a vertex v of $0_\alpha$ if, and only if, side (v, p) belongs to the graph obtained after a previous step of generation e $u \in f(p)$.

Definition 3.3

A Web grammar is said to be monotonic (MWG) if all its rules $\alpha \rightarrow \beta$, f are such that $\{\alpha\} \subseteq \{\beta\}$.

Definition 3.4

A Web grammar is said to be context sensitive (CSWG) if for all its rules $\alpha \rightarrow \beta$ , f, it is true that:

1. We can find a vertex $u \in \alpha$ such that $\alpha \setminus u$ is a subgraph of $\beta$
2. when g: $\{\alpha\} \setminus \{u\} \rightarrow \{\beta\}$ is an isomorphic inclusion, we have $f(v) \supseteq g(v)$ if $v \in (\{\alpha\} \setminus \{u\})$.

Definition 3.5

A CSWG is said to be context free (CFWG) if in all its rules the left side consists of one vertex only.

Definition 3.6

A CFWG is called linear grammar (LWG) if the right side of all its rules and I's graphs do not contain more than one vertex with an index belonging to $V \setminus \Sigma$

Definition 3.7

A Web grammar is said to be normal (NWG) if in the algorithm of its rules, the correspondence f is such that any vertex of $\alpha$ has no more than one image in $\beta$ and any vertex of $\beta$ has no more than one preimage in $\alpha$ .

For normal Web grammars, the following statement is true: any recursively enumerable set of graphs can be generated with the help of a normal grammar, where in all its rules the number of vertices of the left member, corresponding to the right member through f, does not exceed 4 [6].

Such grammars have been described in literature for the following classes of graphs: separable connected graphs, non-separable connected graphs, Euler's separable and non-separable graphs, planar separable and non-separable connected graphs, planar graphs and all trees.

Now, let a graph be given; several rules can be applied to it, and each of them can be applied to various "places" in the graph, because the graph can contain several identical subgraphs, each of which can be substituted. The same rule can be applied to each graph which has been so obtained, and so on. Such a construction gives rise to a "graph-changing equation" which, in rare cases, (when, e. g., only one substitution rule can be applied to any step) generates a unique trajectory.

We will call the construction, which operates in this way, a transformational Web grammar.

Two Problems in the Description of Graph Dynamics

We consider now not a single trajectory, but, rather, the dynamics of phenomena which we meet in the study of the branching of trajectories.

Consider that a transformational and a generative grammar are simultaneously given, connected in the following way: if the generative grammar $G_1$ is given by the quadruple: $G_1 = < V', \lesssim', P', I' >$, then the transformation grammar $G_2$ is determined by $G_2 = < V^2, \lesssim^2, P^2 >$ where $V^2 \; \lesssim^2 = \lesssim^1$; between $V^2 \setminus \lesssim^2$ there is a one-to-one correspondence, whereas in $P^2$ the rules are included which permit the substitution, in any graph, of the index of each vertex from $V^2 \setminus \lesssim^2$ with the correspondent index from $\lesssim^2$, without any change in the graph's sides.

Generative grammar describes the set of graphs $L(G_1)$ which have been generated by it.

If we apply transformational grammar to some graph $L(G_1)$, we obtain a sequence of graphs; therefore, it allows us to describe the graph dynamics.

If we start from a specific initial graph, transformational grammar provides one of the possible descriptions of operator F (compare it with formula 1.1); we will consider set $M = L(G_1)$ as initial graphs; transformational grammar will give rise to set $N = G_2\ (M)$; the pair of grammars $G_1$ and $G_2$ specifies the graph dynamics.

Our problem is to identify, in N, the subset of graphs having some important properties; as a first example, we will study the set of equilibrium graphs.

According to the language used here, an equilibrium graph is a graph on which no rule of transformational grammar $G_2$ can be applied.

Theorem 3.2

A generative grammar $G_1$ and a transformational grammar $G_2$ having been given,

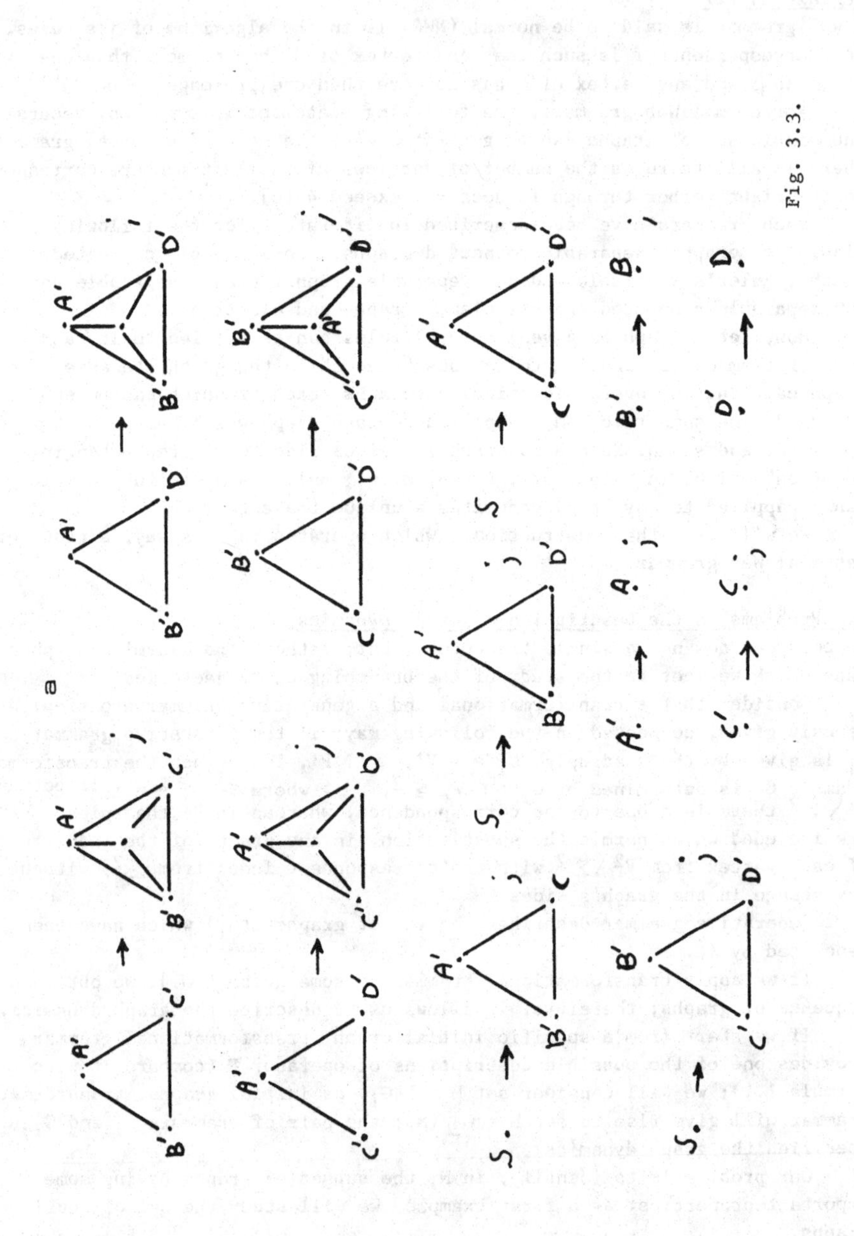

Fig. 3.3.

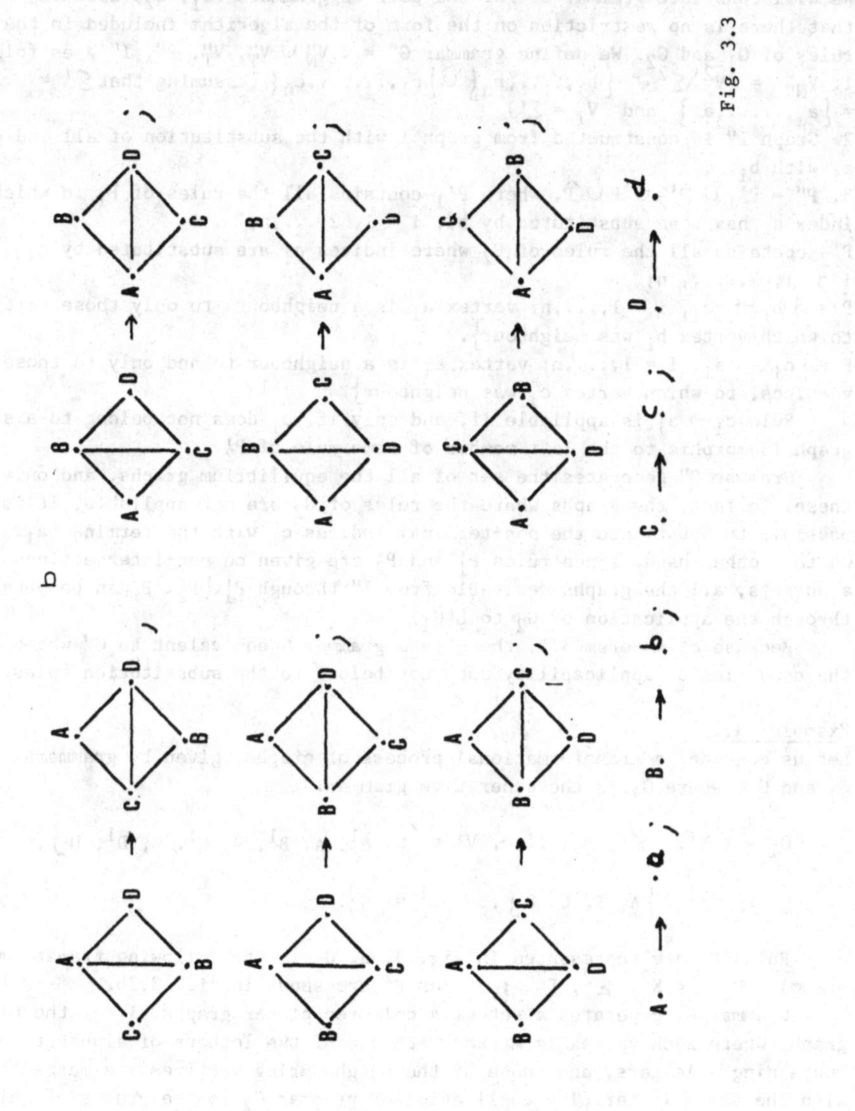


Fig. 3.3

a grammar $G^s$ can be effectively constructed which generates only the equilibrium graphs resulting from the dynamics thus specified.

Proof

We will construct grammar $G^s$ for the pair of grammars $(G_1, G_2)$ assuming that there is no restriction on the form of the algorithm included in the rules of $G_1$ and $G_2$. We define grammar $G'' = < V''_N \cup V''_T, V''_T, P'', I'' >$ as follows:

1. $V_{N''} = V^2 \setminus \Sigma^2 \cup \{b_1, \ldots, b_n\} \cup \{c_1, \ldots, c_n\}$ (assuming that $\Sigma' = \{a_1, \ldots, a_n\}$ and $V''_T = \Sigma'$)
2. Graph $I''$ is constructed from graph I with the substitution of all indices $a_i$ with $b_i$.
3. $P'' = P'_1 \cup P'_2 \cup P \cup \hat{P}$, where $P'_1$ contains all the rules of $P_1$ in which index $a_i$ has been substituted by $b_i$, $i = 1, 2, \ldots, n$;

$P'_2$ contains all the rules of $P_2$ where indices $a_i$ are substituted by $c_i$, $i = 1, 2, \ldots, n$;

$\bar{P} = \{b_i \rightarrow c_i$, $i = 1, \ldots, n$; vertex $a_i$ is a neighbour to only those vertices to which vertex $b_i$ was neighbour$\}$.

$\hat{P} = \{c_i \quad a_i$, $i = 1, \ldots, n$; vertex $a_i$ is a neighbour to and only to those vertices, to which vertex $c_i$ was neighbour$\}$.

Rule $c_i \rightarrow a_i$ is appliable if, and only if, $c_i$ does not belong to a subgraph isomorphic to the left member of some rule of $P'_2$.

Grammar $G''$ generates the set of all the equilibrium graphs, and only these. In fact, the graphs where the rules of $G_2$ are not appliable, it is possible to substitute the non-terminal indices $c_i$ with the terminal $a_i$. On the other hand, since rules $P'_1$ and $P'_2$ are given on non-intersecting alphabets, all the graphs derivable from $I''$ through $P'_1 \cup P'_2 \cup P$ can be obtained through the application of $G_2$ to $L(G_1)$.

Because of theorem 3.1, there is a grammar G equivalent to $G''$ where the condition of applicability does not belong to the substitution rules.

Example 3.2

Let us consider a transformational process of graphs, given by grammars $G_1$ and $G_2$, where $G_1$ is the generative grammar.

$$G_1 = < X', \quad \Sigma', P', I' >, V' = \{S, A^1, A, B^1, B, C^1, C, D^1, D\},$$

$$\Sigma^1 = \{A, B, C, D\}, \qquad I^1 = \{S\}.$$

Rules $P^1$ are represented in Fig. 3.3a, $G_2$ is the following transformational grammar: $G_2 = < X^2, \Sigma^2, P^2 >$; rules $P^2$ are shown in Fig. 3.3b.

Grammar $G_1$ generates a set of 4 coloured planar graphs, i. e. the planar graphs where each vertex is marked with one of the letters of alphabet containing 4 letters, and none of the neighbouring vertices are marked with the same letter. The application of grammar $G_2$ to the graphs of this set, makes it possible to find all the saturated 4-coloured planar graphs. i. e., all the graphs to which no further edge can be added, without destroying the planarity. It is evident that the same set of graphs is the set of equilibrium for the dynamics given by grammars $G_1$ and $G_2$. Grammar $G^s$ is obtained from the union of grammars $G_1$ and $G_2$ which means that $G^s = < V^s, \Sigma^s, P^s, I^s >$ where $V^s = V^1 \cup V^2$, $\Sigma^s = \Sigma^2$, $P^s = P^1 \cup P^2$, $I^s = \{S\}$.

Once the set of equilibrium graphs has been described, we meet another problem, that of describing all the graphs (and not only those generated by grammar $G_1$) which, because of the transformational grammar itself, contain one of the equilibrium graphs generated by $G^s$ at the end of one of the possible branchings of the trajectory.

Theorem 3.3

Generative grammar $G_1$ and transformational grammar $G_2$ having been assigned, a grammar $G^P = < V^P, \Sigma^P, P^P, I^P >$ can be constructed, which specifies all and only those graphs which can be transformed into one of the graphs of $L(G^s)$, through transformational grammar $G_2$.

Proof

We define grammar $G^P = < V_N^P \cup \Sigma^P, \Sigma^P, P^P, I^P >$ as follows:

1. $V_N^P = V_N^s \cup \{d_1, \ldots, d_n\} \cup \{l_1, \ldots, l_n\}$, $\Sigma^P = \Sigma^1$

2. $I^P$ is constructed from $L^s$ with the substitution of all the $a_i$'s with $d_i$'s

3. $P^P = P^{s'} \cup P \cup P' \cup P''$, where $P^{s'}$ contains all the rules of $P^s$ where index $a_i$ is substituted by $d_i$.

$P = \{d_i \rightarrow l_i$, $i = 1, \ldots, n$; vertex $i_i$ is a neighbour to those vertices with which vertex $d_i$ was neighbour$\}$; $P'$ consists of all the rules constructed as follows: in rules of $P_2$ all the $a_i$'s are substituted with $l_i$, and the right and the left sides interchange places. The right side is incorporated in that of the left, so that the application of the correspondent rule of $P_2$ brings the graph back to the initial form.

$P'' = \{l_i \rightarrow a_i$, $i = 1, \ldots, n$; vertex $a_i$ is a neighbour to those and only those vertices with which $l_i$ was neighbour$\}$.

It is easy to verify that grammar $G^P$ is exactly the grammar we were looking for; all the graphs of $L(G^s)$ are to be found in L(GP). Any graph which in one step (i. e. only after one application of the rules of $P_2$) can be transformed into a graph of $L(G^s)$, belongs to $L(G^P)$.

On the other hand, since the rules of $P^s$ and $P'$ operate on non-intersecting alphabets, every graph $L(G^P)$ is either a graph of $L(G^s)$ or arrives at $L(G^s)$ through the substitution rules of $G_2$.

As with the usual problems of dynamics, it is natural to call the set of graphs $L(G^P)$ generated by the grammar of $G^P$, the attraction field of the set of equilibrium graphs generated by grammar $G^s$.

If M is the set of all the initial graphs for the dynamics specified by $G_1$ and $G_2$, and N is the set of graphs obtained by the application of the transformational grammar $G_2$ to the graphs of M, $N = G_2(M)$, whereas $L(G^s)$ is the set containing all the equilibrium graphs of these dynamics, the intersection of set $L(G^P)$ with set M may be non-empty. It is clear that the set $L(G^P)$, the attraction field of $L(G^s)$, contains also $L(G^s)$ itself as a subclass and can also intersect M when $L(G^s)$ is not empty. Sets M, N, $L(G^s)$ and $L(G^P$ are shown in Fig. 3.4

It is also an interesting problem to identify the intersection of the set of all equilibrium graphs with the set of initial graphs M, i. e., to identify all the initial equilibrium graphs.

The solution to this problem is equivalent to that of identifying the set $L(G^s)$.

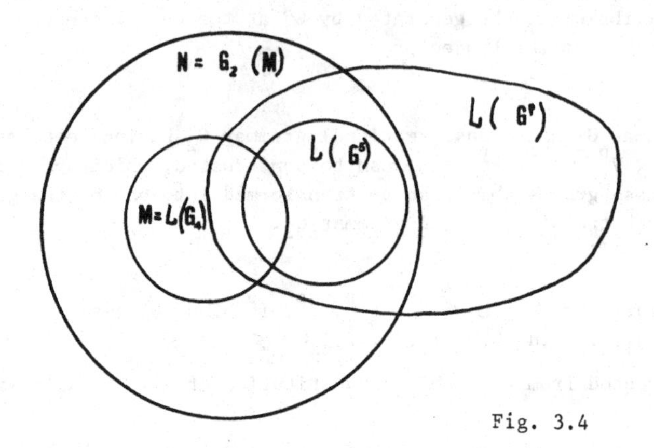


Fig. 3.4

NOTES

1. With substitution $\Psi(\varphi(x))$ we consider all the the values of $\varphi(x)$, $x \in X$ and also those for which $\varphi(x) = x$. Due to this, it is necessary, to perform multiplication (2.3), to consider the isolated vertices of $\Gamma_\varphi$.
2. Successively, $l_\Psi^{j_p}$ and $l_\Psi^{j_s}$ will be called "lengths of vertices": $j_p$ and $j_s$ for short.
3. The set of the vertices of graph $\Gamma$, adjacent to the vertices of its subgraph $\Gamma'$ is called neighbourhood (of ray 1) of $\Gamma'$; the neighbourhood of $\Gamma'$ is denoted by $0_{\Gamma'}$.

REFERENCES


1. Aizermann M. A., Gusev L. A. et al.: 'Dynamic approach to analysis of structures described by graphs (foundation of graph dynamics)', I, II, Automation and Remote Control, 7, 136-151, 9, 123-136, (1977)
2. Gusev L. A., Petrov S. V. et al.: 'Graph substitution operators and dynamics of collegiate hierarchical graphs', I, II, Automation and Remote Control, 1, 113-122, 3, 104-115, (1981)
3. Gusev L. A., Smirnova I. M.: 'Languages grammars and abstract automata models (a survey)' I, II, Automation and Remote Control, 4, 72-94, 5, 73-94, (1968)
4. Petrov S. V.: 'Graph grammars and graphodynamics problem' Automation and Remote Control, 10, 133-138, (1977)
5. Montanari U. G. 'Separable graphs, planar graphs and Web grammars' Inf. and Control, 16, 2, 143-267, (1970)
6. Petrov S. V.: 'Normal form of graph grammars' Automation and Remote Control, 6, 153-157, (1977).

## STRUCTURAL PROPERTIES OF VOTING SYSTEMS

Mark A. Aizerman, Fouad T. Aleskerov
Institute of Control Sciences, Moscow, USSR

## PROBLEM DESCRIPTION

The theory of collective choices [1 - 7] deals with the following typical situation: a group containing a finite but arbitrary number N of individuals (electors) is considered; the group examines a set of variants (which can be plans, projects, appointments to public offices, etc.); each elector can freely have his own opinion as regards the choice of the variant he believes to be the more correct. The problem consists of obtaining a collective decision, starting from individual decisions which do not coincide. The problem can be studied both in a deterministic and probabilistic way. Later, we will discuss only the deterministic formulation of the problem.

This kind of model can be used, as a first approximation, to analyze not only some social processes for which N is sufficiently great (vote theory), but also to analyze the processes for the formation of decisions in "small groups" (in board of directors, committees, Ministries, etc.; in this case, N is higher than 1, but it is not very large and the theory is called "committee theory").

We say that the collective decision has exceeded the condition of "k order majority" if number k has been previously fixed ($1 \leqslant k \leqslant N$) and the decision is adopted by the group if it has been chosen by at least k electors. The well-known different systems of voting, "simple majority" ($k = [N/2] + 1$), "qualified majority" ($k = [\alpha N]$, where $\alpha = 2/3$), "unanimity" ($k = N$), "at least one vote" ($k = 1$), are particular cases of procedure with "k order majority" condition.

Intuitively, we believe that the procedure of "k order majority" is particularly reasonable in cases of simple majority, when $k = [N/2] + 1$; this is based on the idea that "the majority is always in the right" or, in other words, all that is considered to be correct by the majority of the electors is reasonable, therefore, such procedures are valid because they create the conviction of freely taken decisions (each elector exactly contributes by 1/N to the decision). Formal analysis of the problem, however, disagrees with this intuition.

## 2. INDIVIDUAL CHOICE - PROBLEM FORMULATION



*E. R. Caianiello and M. A. Aizerman (eds.), Topics in the General Theory of Structures, 137–148.*
*© 1987 by D. Reidel Publishing Company.*

The problem of collective choice can be exactly formulated as long as the notion of "individual choice" (or elector's action) is not formalized. Such a formalization becomes reasonable only when the choice (both individual and collective) is defined as an algorithm, that is with a regular and effective deterministic procedure.

A is the set of variants (that we consider as finite, for simplicity) and $X \subsetneq A$ its admissible subset. The elector's individual opinion can be characterized in two ways: either assigning his knowledge of the interrelations[1] among the variants of A (or among the subset of A, or among the variants isolated from A) or simply indicating which are the variants he chooses from each non-empty subset $X \subseteq A$.

In the first case, the elector's opinion is formalized as a structure $\sigma$ on the A set. Examples of such formal structures are: binary relations (graphs) on A, introduction of an order in A, different kinds of "hyperrelations" on A, etc. A substructure $\sigma_X$ of the structure $\sigma$ corresponds to each subset $X \subseteq A$. An algorithm of the elector's choice is assigned if we define the structure $\sigma$ on A and some $\pi$ rule. This rule shows how the variants chosen from the subset X can be identified using the substructure $\sigma_X$.

Examples of rule $\pi$ are: extremization procedures if $\sigma$ is a structure assigning an order in A, procedure of extracting predominant variants by means of binary comparisons if $\alpha$ are binary relations among variants, etc.

Thus, if $\sigma$ and $\pi$ are given, for each $X \subseteq A$ the chosen $Y \subseteq X$ set is completely defined, and so also the set of the $\{(X, Y)\}$ couples. The $X \to Y$ correspondence resulting from the choice mechanism $\{\sigma, \pi\}$ represents the choice function $C(\cdot)$, and the set of $\{(X, Y)\}$ couples - its "graph" - represents the description of the choice ("input-output").

In the second case, the electors' opinion is formalized with an individual choice function. The relation between the first and the second description is fixed by the $\pi$ choice rule, an operator which transforms the structure $\sigma$ into the function $C(\cdot)$.

We now introduce three spaces: space $\Sigma$ of $\sigma$ structures, space C of $C(\cdot)$ choice function and space $\Pi$ of $\pi$ rules:

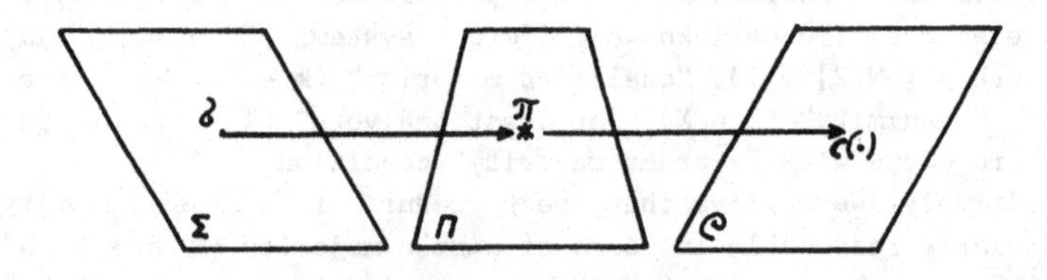


Fig. 1

If we choose the "points" $\sigma$ in $\Sigma$ and $\pi$ in $\Pi$, then the "point" $C(\cdot)$ in C is consequently defined; to the question "how to choose the nth-elector?" we can reply with two different answers: both showing structure $\sigma$ and rule $\pi$, which the elector uses, and directly describing his choice function $C(\cdot) \in C$ (for example, as a table "given X - choose Y" for all $X \subsetneq A$).

The introduction of the formal notion of "structure", "choice rule" and "choice function" transforms the non-formal description gained through the notions of "reasonability" or "motivated choice" into the formal description of some domains in space $\Sigma$ of structures, in space $\Pi$ of rules and in space C of the choice functions (domains $\Sigma_D$, $\Pi_D$ and $C_D$, respectively).

Different points of view on reasonable choice lead to different domains $\Sigma_D$, $\Pi_D$ and $C_D$; however, in each of these cases, these three domains must be "in accordance": each structure $\sigma \in \Sigma_D$ and each rule $\pi \in \Pi_D$ must generate the choice function $C(\cdot) \in C_D$

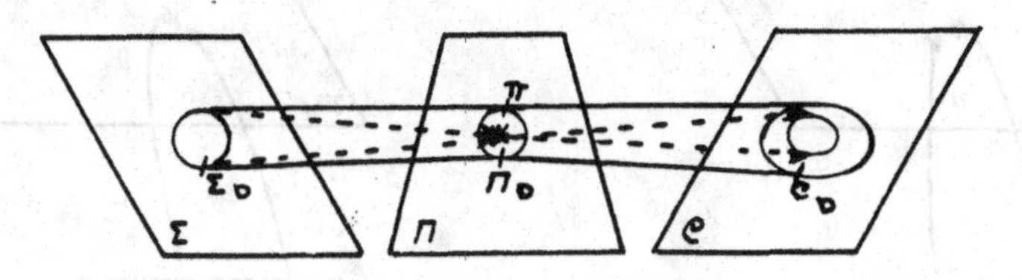

Fig. 2

This "confluence" of different ideas, of reasonable choice, must naturally occur also in the frequent case when the field $\Pi_D$ consists of a single concrete operator $\pi$ (sketched line in Fig. 2).

For the description of domains $\Sigma_D$, $\Pi_D$ and $C_D$ in the theory of choices, a special language is used.

$\Sigma_D$ domains are described by giving the list of the properties of the acceptable classes of structures, for example, if structures provided with a gradation criterion are used, domain $\Sigma_D$ includes "all the gradation criteria", etc.

Structures of the type "binary relations between variants" are normal generalizations of structures provided with gradation criteria; in this case, domain $\Sigma_D$ consists, for example, of "all the acyclic binary relations" or, more particularly, of "all the transitive binary relations", etc.

Domain $C_D$ in the space of choice functions are given through the list of their characteristic properties. Each of these properties reflects some aspects peculiar to the logic of the description ("input-output") of the electors' choice.

Without going deeply into too many details, (see, for example, ref. [8] and references there mentioned) it is worth noting that among these properties, the property of "heritage" (H), "concordance" (C), "elimination of irrelevant variants" (O) and "constance" (K) are the most used. Each of these (and their intersections) determines a correspondent domain $C_D$ denoted by $C_H$, $C_C$, $C_{H \cap C}$, etc.

Standard formulations for $\pi_D$ domains have not been introduced up to now, and some concrete rule or a rule containing "free parameters[2]" are generally used.

## 3. COLLECTIVE CHOICE - PROBLEM FORMULATION

The problem of collective choices is studied according to three different formulations; common to these three formulations is the consideration of the collective choice rule as an operator which makes N "individual opinions" correspond to a "collective opinion". The difference among them consists in the definition of individual and collective opinions or the choice function.

In order to describe these formulations, we introduce three spaces of operators $\pi_I$ $\pi_{II}$ $\pi_{III}$.

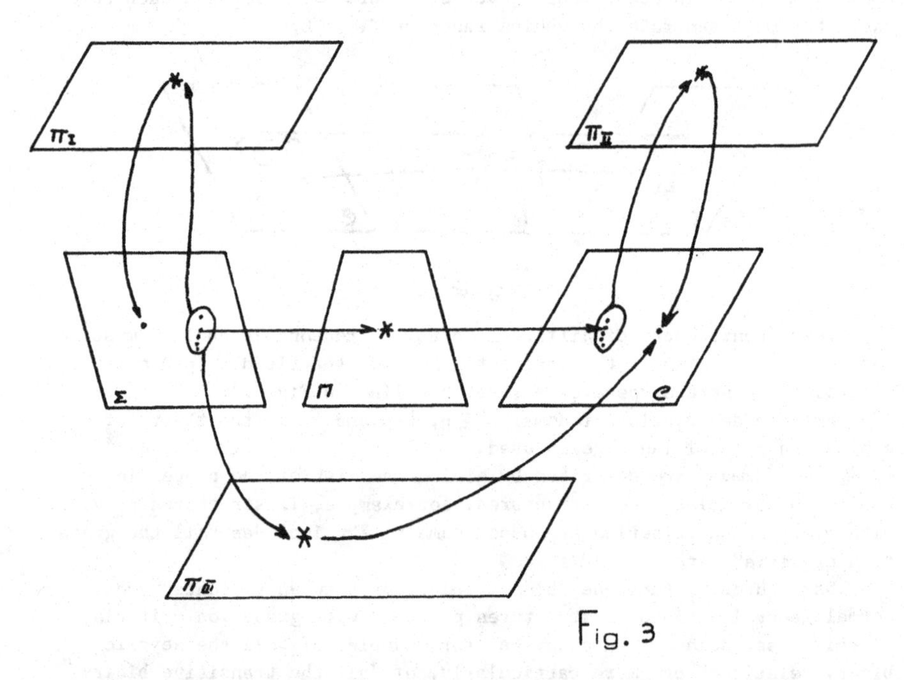

Fig. 3

The elements (or "points") of these spaces are denoted by $\pi_1^*$, $\pi_2^*$ and $\pi_3^*$ and consequently they are referred to as first, second and third type of collective choice operators.

The nth of $\sigma_1, \ldots, \sigma_N$ structures represents the structural profile of the electors and it is denoted by $\{\sigma_i\}$. Similarly, the nth of choice function $C_1(\cdot), \ldots, C_N(\cdot)$ represents the electors' functional profile and it is shown by $\{C_i(\cdot)\}$.

The $\pi_1^* \in \Pi_I$ operator performs a transformation of the structural profile $\{\sigma_i\}$ only in the collective structure $\sigma^*$:

$$\{\sigma_i\} \rightarrow \sigma^*, \sigma^* \in \Sigma$$

In this first formulation of the problem of collective choice, it is implicitly supposed that the choice rule $\pi$ is assigned (see Fig. 1) and that such a rule could be equally applied to any $\sigma_i$, and to the collective structure $\sigma^*$.

The operator $\pi_2^* \in \pi_{II}$ transforms the functional profile of the electors $\{C_i(\cdot)\}$ into the collective choice function $C^*(\cdot)$:

$$\{C_i(\cdot)\} \rightarrow C^*(\cdot)$$

Finally, $\pi_3^* \in \Pi_{III}$ is defined as an extension of the notion of the $\pi$ rule, previously introduced for the description of individual choices: $\pi_3^*$ operator is supposed to be the choice rule, which is to be applied not only to some $\sigma_i$ structures, but also to the profile as a whole and it carries out its transformation in the collective choice function

$$\{\sigma_i\} \longrightarrow C^*(\cdot)$$

In other words, for each $X \subseteq A$, the $\pi_3^* \in \Pi_{III}$ operator builds up the chosen set from the profile $\{\sigma_i\}$, that is, it transforms the $(\{\sigma_i\}, X)$ couple into the collective choice $Y \subseteq X$.

In all the three cases, we consider the spaces of operators satisfying some additional conditions. These conditions are expressed in a similar way for all the three spaces $\pi_1$, $\pi_2$ and $\pi_3$. The main supplementary condition is the location[3] condition.

This location condition is so important and specific for choice systems, that it is convenient to isolate and introduce it in the definition of these spaces, that is, we will study only the spaces $\Pi_I$, $\Pi_{II}$ and $\Pi_{III}$ consisting of local operators $\pi_1^*$, $\pi_2^*$ and $\pi_3^*$.

With an example, we will now show the location condition for space $\Pi_I$, when structure $\sigma$ is in some binary $\beta$ relation on A. In this case, the question of the validity of the relation $x \beta^* y$ for the binary relation $\beta^*$ is decided only by the operator $\pi_1^* \in \Pi_I$ dependent on what is known about the relations between the electors and x and y variants (i. e. whether or not $x \beta_i y$ is true for each elector). In other words, when we solve the problem "Is $x \beta^* y$ true?" we do not consider the information about the electors' opinion in relation to other couples of variants. For example, for the operator corresponding to the unanimity rule, we find that $x \beta^* y$ is true, if $x \beta_i y$ is true for the electors $i = 1,\ldots,N$.

For space $\Pi_{II}$, the $\pi_2^*$ operator "decides" if variant $Y \in X$ is to be included in the collective choice, by using only the information about the inclusion or exclusion variant y in the individual choice.

There are different definitions [9, 10] of the location condition for the operators $\pi_3^*$ of $\Pi_{III}$; for example, we say that $\pi_3^*$ operator is local if variant y included in the choice by x, is dominating as regards binary relations of subsets of electors.

Later, we will consider only spaces of local operators. In each space some subclasses of local operators satisfying any "reasonableness" and "naturalness" condition can be fixed.

Classes corresponding to $\Pi_I$, $\Pi_{II}$ and $\Pi_{III}$ are denoted by $\Pi_{ID}$, $\Pi_{IID}$ and $\Pi_{IIID}$. In all the three spaces D-type classes are introduced in a similar way: four main conditions (operators' characteristic properties) are assigned; each of them isolates some classes in $\Pi_i$; the intersection of these classes in each space $\Pi_i$ (i = I, II, III) leads to $\Pi_{iD}$.

A typical example of the four conditions[4] for the operators belonging to $\Pi_I$ (second column), to $\Pi_{II}$ (third column) and to $\Pi_{III}$ (fourth column) is shown in tab. 1.

For concreteness, the conditions $\Pi_I$ and $\Pi_{III}$ in tab. 1 are written for the case when  is the domain of all the binary relations on A. These conditions can be reformulated in the case where we consider more complex structures, hyperrelations, for example.

All these four relations are equally necessary for each "reasonable" voting operator.

Let us consider now any column in tab. 1. The condition written in each line of this column determines some region in space $\Pi_i$ which corresponds to this column. The intersection of the first and third classes is denoted as "Central Region", whereas the region obtained from the intersection of the Central Region with the fourth class is denoted as "Symmetrically Central Region".

Thus in these three spaces of operators, it is possible to define regions to which "reasonable" voting operators belong.

We now consider another way to define the classes of operators $\Pi^*_i$ (i = 1, 2, 3). We have previously said that "reasonable" considerations lead to the definition of the classes $\Sigma_D$ in $\Sigma$ and $C_D$ in C. If we assume that all the electors have their structures $\sigma_i$ or $C_i(\cdot)$ in such classes, it is natural to ask if the collective decision $\sigma^*$ or $C^*(\cdot)$ belongs to these classes; this is the only case in which the collective decision is as reasonable as the individual ones.

These ideas lead to the notion of "range" ($Q_r$) and specification domain of voting operators[5].

It is natural to consider as classes $Q_s$ and $Q_r$ the regions $\Sigma_D$ and $C_D$ previously introduced. For operators of the first type $\pi^*_1$ the range and the specification domain $Q_r$ and $Q_s$ are subregions of space $\Sigma$ ; for the operators $\pi^*_3$ the classes $Q_r$ and $Q_s$ satisfy the following conditions: $Q_s \subseteq \Sigma$ and $Q_r \subseteq C$; finally, for $\pi^*_2$, $Q_r$ and $Q_s$ are subclasses of C. Evidently, arbitrary operators $\pi^*_i$ (i = 1, 2, 3) cannot be used because in the space of operators, the $S_i$ classes including operators which perform the correspondence

$$Q^i_s \longrightarrow Q^i_r$$

are fixed in a natural way.

These classes $S_i$ (i = I, II, III) are shown as <u>restriction classes</u> on the correspondent $\pi^*$ operators[6].

In the theory of collective choice, we can distinguish two kinds of problems:

<u>Problems of Analysis</u>

Assigning the collective choice rule, i. e. the $\pi^*_i$ operator, it must be determined which characteristic conditions 1 - 4 of tab. 1 are satisfied by the $\pi^*_i$ operator. In addition, if we fix in some way the "rational" domains (for the individual choice) $\Sigma$ or C, then it must be determined whether the assigned rule reconstructs the collective decision in this "rational" domain.

<u>Problems of Synthesis</u>

Once having somehow assigned the domains $Q^i_s$ and $Q^i_r$ (i = I, II, III), it is necessary to find in the corresponding space of the operators

($\Pi_I$, or $\Pi_{II}$ or $\Pi_{III}$) an operator $\pi_i^*$ belonging to the intersection of the Central or Symmetrically Central Region with the class of the restriction $S_i$ defined by domains $Q_s^i$ and $Q_r^i$.

It is particularly important to consider domains $Q_s^i$ and $Q_r^i$ which show requirements of "reasonableness" or "rationality" in individual and collective opinions.

When these two important problems have been solved, two further problems remain unsolved; we will deal only briefly with them.

1. The problem of the defence from electors' manipulation.

It is possible for the elector to manipulate the result of the collective choice using a false choice function (not his true one) to obtain better results for himself, as regards the case in which he "says the truth".

Which operators are immune from such manipulations?

2. The problem of the defence from possible manipulations of the voting organizer. Could the organizer influence the result changing, for instance, the order of the day or the succession in the presentation, even if the voting rule and the elecetors' individual choices remain unaltered?

The dynamic aspects of the vote are strictly connected to this problem, i. e., the study of trajectories arising when voting sequences develop with the same rules and the same conditions of elector's choice.

## 5. MAIN RESULTS

Using the notions introduced, we will study the operators of spaces $\Pi_I$ and $\Pi_{II}$ and only at the end of the work we will mention the results valid for the operators of space $\pi_{III}$[7].

Let us consider now the operators that we have previously called "of k order majority". These operators can be defined both in space $\Pi_I$ and space $\pi_{II}$. In any case ($\Pi_I$ or $\Pi_{II}$) and for any k, it is easy to verify that all the four characteristic conditions 1 - 4 are satisfied, so that the operators "of k order majority" belong to their Symmetrically Central Region both in $\Pi_I$ and $\Pi_{II}$. However, another proposition is also true: "k - majority" operators completely "fill" the Symmetrically Central Region in these spaces.

This statement implies that for any $\pi_1^* \in \Pi_I$ or $\pi_2^* \in \Pi_{II}$ satisfying conditions 1 - 4, there is a k number such that the "k - majority" operator makes an exit equal to that of $\pi_1^*$ or $\pi_2^*$ operators for the same profile correspond to each entry profile.

Thus, if we are limited in the choice of the rule by conditions 1 - 4, usual rules, such as "simple majority" ($k = [N/2] + 1$), or "qualified majority" ($k = [2/3\ N]$) rules, etc., we will not meet objections.

Let us consider now "k - majority" operators in $\Pi_{II}$ from another point of view: let us choose the domain $H \cap C \cap O \equiv Q_s$ in C, and demand that $A_s = Q_r$[8]. It has been demonstrated [11] that there are no "k - majority" operators satisfying this requirement. In addition, if we consider a wider domain $Q_s = H \cap C$ (or $Q_s = H \cap O$) then in $\Pi_{II}$ there is only one "k - majority" operator that satisfies this condition. It is the extremal "unanimity" operator (with $k = N$) for $Q_s = H \cap C$ or the operator of 1 order majority ($k = 1$) for $Q_s = H \cap O$.

The situation is similar in space $\Pi_I$: if in $\Sigma$, $Q_s$ is the domain of all the binary transitive relations, and if the equality $Q_s = Q_r$, is valid, then, among all the "k - majority" operators only the "unanimity" operator (with $k = N$) does not satisfy this condition [9].

These results show the intrinsic contradictions in the practical use of the "k order majority" rules: these operators give as results objects (structures or choice functions which do not satisfy the same conditions compulsory for each isolated elector[10].

If, instead of the Symmetrically Central Region, we consider the Central Region in $\Pi_I$ and $\Pi_{II}$ (fixed by conditions 1 - 3), in it there are complete classes of operators having non-empty intersections with the correspondent Central Region: for $\Pi_I$ these complete classes are defined as regards the domain of all the acyclic transitive relations in $\Sigma$ and for $\Pi_{III}$ as regards the domain $H \cap C \cap O$ in C. But these intersection operators cannot be representded as "k - majority" operators for some fixed k, consequently, such operators are not neutral with respect to the electors [11] (condition 4 is not satisfied).

If we look for the operators satisfying the restrictions on the range and on the specification domain, mentioned before, and if conditions 1 - 2 are considered undoubtly necessary, then neutrality conditions related to the variants (3) and related to the electors are irrelevant both in $\Pi_I$ and $\Pi_{II}$. This conclusion has been called the principle of reciprocally exclusive neutralities.

What is the sense in eliminating condition 4? - What does the neutrality related to the electors mean? A great number of publications, going back to K. Arrow's classical monography, makes it possible to answer exactly this problem.

It results that for each operator $\pi_1^*$ (or $\pi_2^*$) belonging to the Central Region, the totality of the "privileged" groups of electors can be fixed in such a way that the collective choice specified by this operator always coincides with the union of the decisions taken by the unanimity of each group.

The situation is also admissible in which some electors do not belong to any of these groups, so that such electors do not effectively influence the collective decision. Operators of this kind have been called [11, 12, 14] federative.

It has been demonstrated that in the Central Region CR of $\Pi_I$ and $\Pi_{II}$ spaces, two subregions CR1 and CR2 can be isolated, which satisfy the following inclusion property:

$$CR2 < CR1 < CR$$

In this case, for the operators belonging to CR1 the federation is defined by the privileged group of electors ("oligarchy"), and for the operators of CR2 by a single elector ("determined elector" or "dictator"). It also results that CR1 (or CR2) is exactly coincident in $\Pi_I$ with the class of operators with respect to which the domain of all the binary transitive relations (strongly transitive, respectively) is closed in $\Sigma$ .

In space $\Pi_{II}$ of oligarchy operators, that is of operators of CR1, the class of restrictions S, for which $Q_s \subseteq A_r = H \cap C$, is defined, and region CR2 ("dictators" operators) involves all the operators to which domain $H \cap C \cap O$ is closed in C. Remember now that all the choice mechanisms generate in C all the functions of the domain $H \cap C \cap O$ and in $\Sigma$ these mechanisms are correlated to transitive functions. Thus the two formulations of the problem of collective choices (in spaces $\Pi_I$ and $\Pi_{II}$) give rise to collective and individual choices, which are compatible only when the equality of the elector is "greatly" disturbed; this logic is attributed only to the $\pi_2^*$ operators of CR1 ("oligarchy" or "dictatorship"); and only for $\pi_2^*$ operators of CR2 ("dictatorship").

Thus, for the spaces $\Pi_I$ and $\Pi_{II}$ the problem is completely solved. However, another situation develops in space $\Pi_{III}$. Here difficulties arise from the fact that for $\pi_3^*$ operators, the range $Q_2$ and the specification domain $Q_s$ belong to different sets. $Q_r$ belongs to the set C and $Q_s$ to $\Sigma$.

In most studies dealing with this space, this difficulty is avoided assuming that choice functions $C(\cdot) \in C$ are generated by the extremization rule. In these cases, the problem is not reduced to the analysis of space $\Pi_{III}$, but to the analysis of region CR1 of $\Pi_I$.

It is interesting to consider the specific characteristics of the operators of $\Pi_{III}$; but the general method for the elaboration of space $\Pi_{III}$ has not been developed yet. Open problems of the collective choice theory cover both theoretical and practical aspects.

From a theoretical point of view, to speak of completeness of collective choice theory, it is necessary to study the operators of $\Pi_{III}$ in detail. For this purpose, it is necessary to introduce the notion of location in $\Pi_{III}$ and the list of conditions analogous to 1 - 4 in such a form that all the three spaces $\Pi_I$, $\Pi_{II}$ and $\Pi_{III}$ should be "jointed" so that the successive application of operators $\pi_1^*$ and $\pi_3^*$ of the Central Region in $\Pi_I$ and $\Pi_{III}$ gives rise to $\pi_2^*$ operators of the Central Region in space $\Pi_{II}$.

From a practical point of view, these problems involve a search for reasonable and non-contradictory voting operators.

## 6. PROBLEM STORY

As it often occurs, the sequence of events in the development of ideas, which are the basis of collective choice theory, differs from the one presented in review articles. In collective choice theory, all has started from merely applicative problems. The discussion on the way to reach collective decision goes back to ancient times, it has more than 200 years of history. The scientific analysis of the problem began with the monography by the French Science Academician Borda [15] at the end of the XVIII century. New Academicians were elected with one of the variants "of k - majority" to the Academy and Borda drew the other Academicians' attention to the defects of this rule and he suggested a different procedure for the election. Electors had not to choose their favorite candidate, but they had to list the candidates according to their preference. After this gradation procedure, the choice was made with an extremization process. As a consequence of Borda's monography, the Academy changed the mechanism for the election; later, in the time of the first Empire, Napoleon tried to award the Academic title to his protegee, and as he did not succeed in this, he restored the previous procedure.

The introduction of what we have previously called $\sum$ set and space $\Pi_I$ is due to Borda's monography.

The problem was later tested again, during the Convention period; during the meetings of the Convention, decisions of vital importance for the fate of the revolution had to be taken. Some members of the Convention were famous scientists; one of them, Condorcet, continued the analysis started by Borda. In particular, he demonstrated that the problem of constructing a collective order from the set of individual orders was without a solution if "k - majority" operators were used and if on the basis of the solution method the two by two comparison of the classification was adopted. In fact, he showed that if the electors present their opinion with a precise classification, then the "k - majority" operator does not frequently lead to a collective order, but to a cycle, therefore, the final collective choice cannot be obtained by a two-by-two comparison.

In the XIX and in the first half of the XX centuries, other scientists went back to the analysis of voting systems and attempts to cope with the difficulties appearing in Borda's and Condorcet's works failed: if the suggested methods were local, that is if the decision on the collective rank of two variants depended only on their individual rank and if the "k - majority" rule was used, then the same difficulties met by the founders of the theory appeared.

These unsuccessful attempts have lead, at the end of our century, to the change of the formulation of the problem itself: a passage has taken place from the analysis of the procedures to that of their synthesis.

K. Arrow [13] in his principal monography on this topic advanced (in the terms still used) the problem of the analysis of the Central Region in the space $\Pi_I$ and found that if the additional closure condition of binary transitive relations was imposed, then, inevitably, the existence of a "dictator" was reached. This conclusion was so unexpected, that it was called "Arrow's paradox", and from this several works resulted.

Still now, these studies continue and the results obtained have been summarized in the monography [1 - 3] and in review articles [4 - 6].

In these hundred or so works, the study is limited only to the Central Region and to the Symmetrically Central Region, where the S restriction classes of $\pi_1^*$ operators are specifically limited.

A. Sen [1, 4] attracted the scientists' attention because for the first time he pointed out that space $\Pi_I$ does not deal with problems of transformation of structures, so that he advanced [1] the problem of collective choice in space $\Pi_{III}$. We have already mentioned before the difficulties which can arise when correspondent operators are analyzed. As a consequence, the "revolution" caused by A. Sen in this field could not lead to new developments in successive works.

Only recently [11], $\pi_2^*$ operators have been introduced and this leads to the hope of arriving at a complete theory. The construction of this theory will be dealt with in future works.

NOTES:

1. The idea of mutual relations between variants is based on the notion of preferability of a variant in comparison with another one, or, of similarity-difference between variants, or of "admissibility" and "interdiction" of a variant with respect to another one, etc.

2. For instance, a criterion is assigned, for which the evaluation differs from the maximal evaluation in X by no more than $\varepsilon$ ( $\varepsilon$ - parameter).

3. The collective choice operators not satisfying the location condition are generally studied in evaluation theory and not in vote theory.

4. In literature, also other kinds of conditions have been studied and theorems have been proved, which establish reciprocal relations between these different relations and those shown in tab. 1.

5. The domain of voting operators is obviously an n-ple given by a direct product of $Q_s \times Q_s \times \ldots\ldots\ldots \times Q_s$ type, where $Q_s \leq \Sigma$ or $Q_s \leq C$, as operators $\pi_1^*$, $\pi_3^*$ or $\pi_2^*$.

6. For operators of the first and second classes $Q_s^i$ and $Q_s^i$ belong to the same space $\Sigma$ or C. For these domains, we assume that the inclusion $Q_s^i \leq Q_r^i$ is satisfied; this corresponds to the natural idea that opinions acceptable as individual are acceptable as collective (and vice versa, in the case of the equality $Q_s^i = Q_r^i = Q$.

7. Most of the publications on this topic (more than 500 have been reported in the review of 1980) consider operators of $\pi_1^*$ type; operators of $\pi_3^*$ are studied in relatively fewer articles, and, apparently, only in one work [11] are the $\pi_2^*$ type operators considered.

8. Remember that this is a "reasonable" domain.

9. There is an exception when the number of variants is equal to two and the number of operators is odd. In this case, as demonstrated by K. Arrow [13], the above mentioned difficulties can be avoided. Elsewhere [12], it has been demonstrated that the problem can also be solved in other particular cases (when other relations between the number of variants and the number of electors are satisfied).

10. One can believe that such "contradictions" arise only for some specific profile of the D domain in question. Unfortunately, this is not the case. Profiles for which these "contradictions" arise are as frequent as the profiles for which these rules are very suitable.

11. There is an exception for space $\Pi_I$ - the "unanimity" rule (k = N).

REFERENCES

1. Sen A. K.: Collective Choice and Social Welfare - Holden-day, San Francisco, Cambridge, London, Amsterdam, 215, (1970)
2. Arrow J. S. Kelly: 'Impossibility Theorems' - Academic Press, New York, 194, (1978)
3. Schwartz T.: 'The Logic of Collective Choice' - School of Social Sciences, Univ. of California, 389 (1980)
4. Sen A. K.: 'Social Choice Theory: a Re-examination' - Econometrica, 45, n. 1, 53 - 89 (1973)
5. Plott C. R.: 'Axiomatic Social Choice Theory: an Overview and Interpretation' - Amer. J. of Political Sciences, 20, n. 3, 511 - 546 (1976)
6. Emerlianov S. V., Nappelsaum E. L.: 'Techniceskaia Kibernetica' 10, 120 - 214 (1978)
7. Aizerman M. A.: 'Automatica i Telemechanica', 12, 103 - 118 (1981)
8. Aizerman M. A., Malischevski A. V.: 'Automatica i Telemechanica' 2, 65 - 83 (1981)
9. Ferejohn J. A.Grether D. M.:'Some new Impossibility Theorems' Public Choice, 30, 35 - 42 (1977)
10. Bordes G.: ' On the Possibility of Reasonable Consistent Majoritarian Choice: Some Positive Results' - Economic de l'enviroment. Université de Bordeaux, Faculté des Sciences Economiques, 1 - 23 (1981)
11. Aizerman M. A., Aleskerov F. T.: 'Automatica i Telemechanica' 9, 127 - 151, (1983)
12. Aizerman M. A., Aleskerov F. T.: 'Local Operators in Models of Social Choice' - System and Control Letters, 3, 1 - 6, (1983)
13. Arrow K. J.: ' Social Choice and Individual Values' - Yale Univ. Press, 2-nd ed. (1963)
14. Mirkim B. G.: ' Nauka', 104 - 119 (1979)
15. Borda J. C.:'Mémoire sur les Elections ou Scrutin' - Histoire de l'Académie des Sciences pour 1781, Paris, (1784)
16. Condorcet: 'Esposition des Motifs et des Principes du plan de constitutions' - Ar 12 (1793)
17. Ferjohn J., Fishburn P. C.: 'Representation of Binary Decision Rules by Generalized Decisiveness Structure' - Econ. Theory, 1, 28 - 45, (1979)

# APPLICATION OF PREDICATE CALCULUS TO THE STUDY OF STRUCTURES OF SYSTEMS

S. V. Petrov
Institute of Control Problems
Soviet Academy of Sciences, Moscow, USSR

## 1. STUDY OF THE STRUCTURE ON THE BASIS OF THE DESCRIPTION OF THE SYSTEM

Several works on the general theory of systems [1, 2] suggest studying their structure according to the following scheme:

1. All the data of the system can be described as a set of n-ples of the values of assigned parameters $U = \{A_1, \ldots, A_n\}$, $n < \infty$. The U set, the $A_i$ parameters and the neighbourhoods $dA_i$ are constant characteristics of the system. Each n-ple

   $$\omega = a_{i_1}, \ldots, a_{i_n} \quad (a_{i_j} \in dA_j,\ j = 1, \ldots, n)$$

   represents the state of the system.

   State $a_{i_1}, \ldots a_{i_n}$

   is possible if during the functioning of the system all the values of the parameters $A_1, \ldots, A_n$ simultaneously become equal to

   $$a_{i_1}, \ldots, a_{i_n}.$$

   The system is defined if U, $dA_i$ is given for all the i's and the set of possible states $R \subseteq dA_1 \times \ldots \times dA_n$.
2. For the S system defined by the U-set of the parameters and by the R-set of the possible states, we can consider $2^n$ subsystems; the subsystem S' is defined by the set of the $X \subset U$ parameters and by the $R' = \{x/X \mid x \in R\}$ set of the possible states, where the x/X n-ple is obtained from x, discarding the values that do not belong to the X parameters.
3. The S system formed by S'(X, R') and S"(Y, R") subsystems is defined by the set of $X \cup Y$ parameters and by the set of possible states $\tilde{R} = R' * R'' = \{\omega \mid \omega/X \in R',\ \ \omega/Y \in R''\}$.

The operation introduced to obtain $\tilde{R}$ from R' and R" is associative and commutative. Therefore, we can apply it to any (finite) number of operands and consider the systems containing any number of subsystems.

By this hypothesis, the structure of the system S(U, R) is naturally



*E. R. Caianiello and M. A. Aizerman (eds.), Topics in the General Theory of Structures, 149–156.*
*© 1987 by D. Reidel Publishing Company.*

a re-covering of the U-set, so that

$$R = \underset{j=1,k}{*} R_{X_j} \tag{1}$$

where $R_{X_j}$ is the set of possible states of the $S_j$ subsystem, which has been defined by the set of $X_j$ parameters, in accordance with point 2. This re-covering defines the union of the parts of the system. Their states permit the re-building of the set of states of the global system.

The first assumption entails a uniformity of all the states of the system: all have the same characteristics and differ only in their specific values. In general, this hypothesis is not always true; but it is an obvious consequence of the second assumption: it is enough to consider as a system the subsystem described by a part of the parameters used for the description of the whole system. The second assumption implies that different parameters should describe different parts of the system and they should not be general characteristics (as "weight", "cost", etc.). If this requirement is satisfied and each parameter describes an element of the system, then, all its subsystems are determined by different sub-sets of the set of parameters and, since the content of the system is always constant, the first hypothesis is satisfied.

As (1) describes in reality the structure of parameters and not that of the initial system, the connection between the parameters and the elements of the system (element - group of parameters) becomes evident.

The second and third assumptions fix the method for the choice and connection among subsystems. Each subsystem may be only in those states where it appears while working within the whole system. The equality of the common parameters' values establishes the connection among the subsystems.

According to the second hypothesis, the minimal set not respecting (1) is the set of the states of the subsystems. This choice assures that the sets of states of different subsystems are independent.

Any element in the system may be in a single state at every moment. This statement corresponds to the third hypothesis as soon as the initial system is replaced by the system of parameters. The third hypothesis entails that any system, built with S subsystems and described by a U-set of parameters, does not hold more than $||U||$ elements, since each parameter (= element) appears only once. Using many repetitions of a parameter, the class of systems constructed by the initial one and the class of properties expressed by type-(1) equalities can be enlarged.

The functional connection between the values of two groups of parameters can represent an example of such a property. Formal implementation of the third hypothesis can be obtained giving the repetitions new names. Thus, the use of repetitions offers the possibility of joining the subsystem not only through all the common elements, but also through fixed sets of common elements.

The operator is on the set of the states of the system indirectly introduced by the second and third hypotheses are called projection and union; they are the well-known operations of relational algebra.

The set of the states of the system (set of n-ples with a fixed n length) is an n-ary relationship which can be determined by a predicate depending on n variables. Using this connection, previous statements about the associative and commutative power of the union among subsystems (third hypothesis) and the conditions which permit the representation of functional connections between the elements of the system with the aid of repetitions of parameters, can be easily verified.

All this can be confirmed by the following arguments.

As the study of the structure of a system is in reality the study of the mutual action of the parts of the system, any approach implies, either directly or indirectly, separation and connection operations of the subsystems. As we generally deal with the description of the system rather than with the system itself, it is necessary to determine how the description is used and to consider the operations performed on the descriptions. These operations must permit, from the description of the system, the description of its subsystems and, from the different descriptions of subsystems, that of the initial system. Therefore, the success of the chosen approach is determined by the choice of the description method and of the operations. It depends on the level of efficiency exerted on a class of systems and on the kind of mathematical problems one has.

Hypotheses 1 - 3 make it possible to limit this study to the derivation of the consequences in the calculus of first order of the set of the given formulae. All the formulae (both given and derived) belong to a limited class defined by the applicable operations. Finally, we will give a certain number of examples in which the structure of the systems described in accordance with hypotheses 1 - 3 does not contradict usual notions.

<u>Example 1.1</u> A finite state (deterministic) system can be defined as a system described by four parameters: C (moment state), I (input), C' (next state), O (output). Each parameter has a finite set of values. This system has the following structure:

$$(\{C, I, C'\}, \{C, I, O\}).$$

The present definition shows that in automata with finite states the values of C' and O are inteconnected with I and C.

<u>Example 1.2</u> Nets of automata and functional elements can be described by different sets of parameters, however no description makes it possible to reconstruct the behaviour (set of states) of the net from the behaviour of its elements (interconnections between elements are not given). The following statements are simple to verify.

In any acyclic net of functional elements in which each parameter corresponds to a connection between two elements, the set of states may be reconstructed from the set of states of the single elements. This is true for every set of finite states automata, if all the states in all the automata are initial.

So far, we have supposed the delays in the elements of the net equal to zero. This statement would be true for system with cycles if all delays were equal and if two parameters were corresonding to each connection between two elements: one parameter to describe the value of the signal at one moment, and another for the previous one.

In example 1.2, we have assigned data for the description of nets of automata with a system of parameters. We do not know whether or not there are studies on this problem or on the analogous problem of asynchronous nets (for instance, Petri's nets). In ref. [1], the system formed by a processor and three channels is taken as an exmple. The parameters corresponding to each element may have two values, L and H. The system is divided into two subsystems, with the state table.The description of the system and of the subsystems is constructed according to hypotheses 1 - 3.

Previous examples show that the class of systems satisfying hypotheses 1 - 3 is not empty and it contains some systems important in practice. One should note that each method for the study of structure, except the experimental one, is in reality a method for the study of the data on the system and not of the system.

A method for the study of the structures of data consists in the construction of bases of relational data. In the bases of relational data, the connection among data is assigned as a multiple relation. Problems dealing with the structure of data systems are known to be problems of logic. Literature on this matter is massive (see, for example, review article [3]). In the theory of data bases, the parameter is called attribute.

The representation of multiple relations through the union of its projections is the first problem connected with the study of structure under the hypotheses 1 - 3. This is a combinational problem and it generally has exponential complexity. The answer to the question whether it has a significant representation can be obtained with $t^2$ steps, where t is the capacity of the relation [4]. The method for the search of a common representation of the relations of an assigned class depends on the way it has been determined. The class of relations may result from the number of the known common structures. In this case, the problem is reduced to the search for other structures which are valid for this class of relations.

This is the problem of deriving logical consequences from a given set of formulae. As the number of assigned or derived formulae is rather limited, it is better to use special derivation rules instead of the usual rules of first order calculus. In ref. [5], some systems of rules suitable for some classes of formulae are given. They describe the behaviour of a subsystem through that of their subsets and permit a comparison of sets of states of different systems constructed from the initial one in accordance with hypotheses 1 - 3. All the properties fixed by these formulae are naturally called structural properties of the system.

## 2. THE PROBLEM OF THE EQUIVALENCE OF STRUCTURAL PROPERTIES AND THE USE OF HYPERGRAPHS FOR THEIR EXPRESSION

Let us consider now the problem of deriving logical consequences and looking for the equivalences for the formulae of first order calculus of the following type:

$$\forall X \, (\sigma_1 (X) \Rightarrow \sigma_2 (X)) \tag{2}$$

where $X = \{x_1, \ldots, x_k\}$ is the list of variables; $\forall X$ is used instead of $\forall x_1 \ldots\ldots \forall x_k$; $\sigma_1(X)$ and $\sigma_2(X)$ are correctly constructed formulae with a single symbol P of n-ary predicate, with operation & and quantifier $\exists$ on the variables. Formulae $\sigma_1(X)$ and $\sigma_2(X)$ can contain the equality predicate ($x = y$) if x and y are considered equivalent in all cases of predicate P. All the variables belong to the infinite list of variables U. X is the set of all the independent variables in formula $\sigma_1(X)$.

Formula (2) depends on the single symbol of P predicate, and thus it expresses the properties of n-ary relations (truth tables of n-ary predicates). These properties can be described approximately in the following way. From the initial n-ary relation, we can build a certain number of other relations applying projection and union (equivalent to operations $\exists$ and &). The expressions $\sigma_1(X)$ and $\sigma_2(X)$ describe two such processes, whereas formula (2) compares the results: it is true when the relationship constructed according to $\sigma_1(X)$ is included in the relationship constructed according to $\sigma_2(X)$. In particular, formula (2) determines the property of reconstructability of the given projection relation from a certain set of other projections when

$$\sigma_2(X) = \exists (Y - X) \, P(X)$$

In the theory of data bases, the properties expressed by formula (2) are called dependencies; here they are called structural properties.

The research of dependencies logically resulting from those given leads to the axiomatization of certain classes of dependencies. The solution of this problem (the construction of the system of rules) is based on the reconstruction process of counterexamples of the statement regarding the chosen logical consequence. This process is a special case of the general one of the search for the solution for an assigned set of disjointed dependencies. In this special case, the problem makes it possible to find all the consequences. In ref. [5], the cases where this process is effective are enumerated, and the complexity of the process for different classes of formulae is evaluated.

When it is necessary to derive the consequences, the way of expressing formula (2) is almost irrelevant; but it is essential for the problem of the search for equivalent formulae, as the existence of equivalent but syntactically different formulae complicates the problem. In this case, the following way of expressing formula (2) is convenient.

The (V, W) couple, where $W \subseteq P(V) - \phi$ and P(V) is the boolean of the finite set V, is called hypergraph. Let us connect the hypergraph $\Gamma_\sigma$ with the expression in (2). For this purpose, we must transform the expression $\sigma$ in the following way:

$$(\exists \, y_1 \; P(X_1) \, \& \ldots \& \; \exists \; y_t \; P(X_t)) \tag{3}$$

if in $\sigma$ there is no equality predicate, the transformation can be carried out giving the variables a new name and eliminating the parentheses.

If $\sigma$ contains equalities, we replace the n-ary predicate P with an (n+1)-ary predicate

$$\tilde{P}(x_1,\ldots, x_n, x_{n+1},\ldots, x_{n+\ell}) \equiv P(x_1,\ldots, x_n) \,\&\, (x_{i_1} = x_{n+1}) \,\&\, \ldots \,\&\, (x_{i_\ell} = x_{n+\ell})$$

where $(x_{i_s} = x_{n+s})$ are equalities resulting in $\sigma$. We will call the variable $x_{n+s}$ the copy of the variable $x_{i_s}$ in predicate P. This predicate always exists and it allows each equality in $\sigma$ to be replaced with the expression

$$\exists (y - \{x_{i_s}, x_{n+s}\})\ \tilde{P}(y)$$

where y is the list of the free variables of $\tilde{P}$.

X is the set of the free variables of $\sigma$. The set of the V vertex of $\Gamma_\sigma$ is defined through a one-to-one correspondence

$$\varphi : X \cup Y \rightarrow V.$$

For each union element $\exists y_i\ \tilde{P}(x_i)$ a hyperedge

$$w_i = \left\{ \varphi(x_{i_{1_1}}),\ldots, \varphi(x_{i_{1_i}}) \right\}, \quad \text{where } (x_i \setminus y_i) = \{x_{i_1},\ldots,x_{i_{1_i}}\}$$

is introduced in W.

The set consists exclusively of all these hyperedges.

$\Gamma_\sigma$'s vertices are classified by the alphabet $A = \{A_1,\ldots, A_n\}$ $\Psi : V \rightarrow A$ so that $\Psi(G) = A_i$ if $\varphi^{-1}(G)$ occupies the nth place in P or if it is a copy of the nth variable of P.

The couple of hypergraphs

$(\Gamma_{\sigma_1}, \Gamma_{\sigma_2})$ constructed for $\sigma_1$ and $\sigma_2$ corresponds to formula (2) if $V_1 \cap V_2 = \varphi_1(X) = \varphi_2(X)$. It is clear that for each

$$G \in V_1 \cap V_2, \quad \Psi_1(G) = \Psi_2(G).$$

In the system of derivation rules in ref. [5] there are two systems which can be easily translated into the chosen language and which are suitable for generating all the consequences from the set of the couples

$$(\Gamma_{\sigma_1}, \Gamma_{\sigma_2})_j, \quad j = 1, 2, \ldots, S.$$

DEFINITION 2.1

Given $\Gamma_1 = (V_1, W_1, \Psi_1)$ and $\Gamma_2 (V_2, W_2, \Psi_2)$, the correspondence $h : V_1 \rightarrow V_2$ is called homomorphism if

1. $\forall G \in V_1 \qquad \Psi_1(G) = \Psi_2(h(G))$.

2. $\forall \omega_1 \in W_1 \quad \exists\, \omega_2 \in W_2 : \left\{ h(G) / \omega \in W_1 \right\} \subseteq \omega_2$

<u>Lemma 2.1</u> The couple $(\Gamma_1, \Gamma_2)$ is tautological if and only if there is a homomorphism

$$h : V_1 \rightarrow V_2 \text{ with } h(V_1 \cap V_2) \subseteq V_1 \cap V_2$$

We will not give the proof here, because it is easily reconstructed by the proof in lemma 2.6, given in ref. [5], substituting the formulae with the hypergraphs.

The use of homomorphism makes it possible to show that the implication of formula (2) is always "the transitive closure" of the implication of many formulae.

Lemma 2.2 The couple $(\Gamma_1, \Gamma_2)$ entails $(\Gamma_3, \Gamma_4)$ if, and only if,
1. There are homomorphisms $h_1$ and $h_2$ such that the couples $(\Gamma''_3, \Gamma_1)$ and $(\Gamma_2, \Gamma''_4)$ satisfy the conditions of lemma 2.1.
2. The couple $(\Gamma'_3, \Gamma'_4)$ is tautological, where

$$\Gamma_3 = \Gamma'_3 \cup \Gamma''_3 , \quad \Gamma_4 = \Gamma'_4 \cup \Gamma''_4$$

Also the proof of this lemma can be easily derived from the proof of the theorem 4.3 shown in ref. [5].

From lemma 2.2, it follows that the criterion for the existence of couples equivalent to the couple $(\Gamma_1, \Gamma_2)$ is represented by the existence of homomorphical re-coverings on the hypergraphs $\Gamma_1$ and $\Gamma_2$. As the indices of the vertices do not change through homomorphism, the following theorem is true:

Theorem 2.3 The complexity of the proof of equivalence depends polynomially on the number of vertices of the hypergraph in the formula. It is connected with the structure of the parentheses and does not enable us to find and to discard formulae which are equivalent to the given one.

The derivation of structural properties equivalent to those given allows us to pose and solve the problem of reducing the redundancy of a system.

REFERENCES

1. Cavallo R. E., Klir G. J.: 'Reconstructability Analysis of Multi-dimentional Relations: a Theoretical Basis for Computer-Aided Determination of Acceptable System Models' - Int. J. General Systems, 5, 143 - 171 (1979)
2. Conant R. C.: 'Detecting Subsystems of a Complex System' - IEEE on Systems, Man and Cybernetics, SMC-2, 4, 550 - 553, (1979)
3. Gorchinskaya O. Yu.: 'Theoretical Aspects of Relational Data Base Design' - Automation and Remote Control, 44, n. 1, part. 1, 1 - 17 (1983)
4. Gorchinskaya O. Yu., Petrov S.V., Tenenbaum L. A.: 'Decomposition of Relation Schemes and Logical Design of Relational Data Bases' - Automation and Remote Control, 44 I, n. 2, part 2, 269 - 275; II, n. 3, part 2, 399 - 406, (1983)
5. Beeri C., Vardi M. Y.: 'Formal Systems for Tuple and Equality Generating Dependencies' - SIAM J. on COMPUTING, 13, n. 1, 76 - 98, (1984)
6. Yanakakis M., Papadimitriou C. H.: 'Algebraic Dependencies' - J. of Computer and System Sciences, 25, 2 - 41, (1982).

# TOURNAMENT FUNCTIONS IN PROBLEMS OF COLLECTIVE CHOICES

V. I. Vol'skiy
Institute of Control Sciences, Moscow, USSR

In the field of Decision Theory, the main problem is the choice of the best variant of a given set. In real situations, variants can be sectorial development projects, locations of industries, etc.

Each time the variants are assessed on the basis of a set of criteria, or these evaluations are made by a community of individuals, individual evaluations must be aggregated in order to gain a "global preference". The problems of the aggregation of individual opinions (generally represented as binary relations) are indicated as problems of collective choice, while the problem of gathering evaluations which depend on a set of criteria is called a problem of "multi-criterial" choice.

In problems of collective choice, individual evaluations are usually expressed as a set of binary relations $\{G_i\}$, $i = 1, \ldots, n$. In problems of "multi-criterial" choice, they are given as $\varphi_i(x)$ evaluations of $x \in A$ variants in the space of criteria $\{\varphi_i\}$, $i = 1, \ldots, n$. We notice that a problem of collective choice can be reduced in an equivalent way to that of a "multi-criterial" problem, if the binary relations satisfying both the conditions of acyclicity and transitivity, that is, weak order relations, are considered as individual binary relations. Then, each such $G_i$ relation may be put in correspondence with the numerical function $\varphi_i(\cdot)$, so that

$$(x, y) \in G_i \Longleftrightarrow \varphi_i(x) > \varphi_i(y).$$

In practice, from now on, we will consider only the "multi-criterial" formulation of the problem.

In this work, we shall investigate the aggregation procedures of individual evaluations, given on a basis of a set of binary criteria or relations; these procedures use the conception of tournament matrices for whose study the techniques of decision [1] theory [1] are applied.

Let us specify a finite set A of variants $x \in A$. Each non-empty subset of variants $X \subseteq A$ can be used for the choice. The act of choosing consists in the identification of a subset $Y \subseteq X$, according to some $\pi$ rule, from an assigend $X \subseteq A$.

All the various acts of choosing define the set of pair $\{(X, Y)\}$,



*E. R. Caianiello and M. A. Aizerman (eds.), Topics in the General Theory of Structures, 157–162.*
*© 1987 by D. Reidel Publishing Company.*

or choice functions $Y = C(X)$.

By means of the evaluations $\varphi_i(x)$ of variants $x \in A$, we build up a square matrix $|A| \times |A|$, $M_A = \|m_{xy}\|$. We put the number of criteria $\varphi_i$ for which variant y is better than x, in the position given by the intersection of line x with column y.

That is, $M_A$ matrix is a matrix whose elements inform us of the comparison among couples of variants $x \in A$ in the space of criteria $\{\varphi_i\}$, i = = 1, ...., n. Matrix $M_A$ can be considered like a table of the results of a tournament, if variants $x \in A$ are considered as players, while the element $m_{xy}$ is considered as the number of games, in which the player y has won against player $x$.

Matrix $M_A$ is called a tournament matrix in a "multi-criterial" analysis. It differs from the general tournament matrix, because in addition to the relations

$$m_{xy} + m_{yx} = n \quad \forall x, y \in A \ (x \neq y); \quad m_{xx} = 0 \ \forall x \in A,$$

the following triangular inequality is also valid:

$$m_{xy} + m_{yz} \geqslant m_{xz} \qquad \forall x, y, z \in A.$$

Let us consider some construction methods of choice procedures using the tournament matrix $M_A$. We introduce a family with a choice rule parameter

$$\left\{\pi_s, \quad s \in [1, \infty]\right\}$$

on matrix $M_A$:

$$\pi_s \Rightarrow Y_s = \left\{ y \in X \mid D^s_x (y) = \min_{x \in X} D^s_X (x) \right., \text{ where}$$

$$D^s_X (x) = \sqrt[s]{\sum_{z \in X} (m^s_{xz})} \ ;$$

fixing the value $s = s_0$, a concrete choice rule is obtained.

When s = 1, the usual choice rule in a tournament with total score $\pi_{s_1}$ is obtained (that means that variants with the minimal negative score out of all the other variants are chosen).

When $s = \infty$ , we have the Kramer $\pi_{KR}$ rule, introduced in dynamic problems of choice theory

$$\pi_{KR} \Rightarrow Y_{KR} = \left\{ y \in X \mid D^{\infty}_X (y) = \min_{x \in X} D^{\infty}_X (x) \right., \text{ where}$$

$$\left. D^{\infty}_X (x) = \max_{z \in X} m_{xz} \right\},$$

that is, with Kramer's rule those variants are chosen, for which the maximal number of criteria is minimal; so that the variant is loser with respect to each altrnative.

Let us consider now the properties of choice function

$$\left\{ C_s (X), \quad s \in [1, \infty] \right\}$$

which corresponds to the tournament rule

$$\left\{ \pi_s, \quad s \in [1, \infty] \right\}$$

Theorem 1

For each $s \in [1,\infty]$, the inclusion $C_s(X) \subseteq C_{PAR}(X) \; \forall \; X \subseteq A$, where $C_{PAR}(X)$ is the function of Pareto's optimal choice, is valid.

Therefore, choice functions belonging to the $\{ C_{s'}(X), \; s \in [1, \infty] \}$ family can be used to identify Pareto's set.

The solution of this problem is often useful in questions of "multi-criterial" [2] choice.

For the description of choice functions, the following characteristic conditions are generally used (see ref. [1] and relative references).

Definition 1

Choice function C(X) satisfies heritage conditions (H), if $X' \subseteq X$ implies

$$C(X') \supseteq C(X) \cap X';$$

concordance conditions (C), if $X = X' \cup X''$ implies

$$C(X) \supseteq C(X') \cap C(X'');$$

and independence conditions from excluded variants (O), if $C(X) \subseteq X' \subseteq X$ implies

$$C(X') = C(X).$$

We state that choice function C(X) fullfils maintenance choice condition (C M condition) if for $X' \subseteq C(X)$ condition $C(X') = X'$ is satisfied.

Using these conditions, choice functions [1], which are mostly used, can be described. For example, Pareto's choice function satisfies conditions H, C and O.

In practice, however, there are also choice functions which do not satisfy any of these characteristic conditions.

Theorem 2

For each $s \in [1, \infty]$, choice function $C_s(\cdot)$ does not generally satisfy any of the conditions C, H, O and C M.

The problem of finding some charateristic condition arises for the description of these and other choice functions, which are essentially non-classic.

In this work, we introduce characteristic conditions of a new type, assuming that $C(X) \subseteq C^*(X)$, where $C^*(X)$ belongs to a well-known class of choice functions (for example, $C^*(X)$ coincides with the class of Pareto's choice functions $C_{PAR}(X)$).

Having formulated the new characterisic conditions, it is possible to use the notion of superposition of choice functions. The superposition of functions $C(\cdot)$ and $C^*(\cdot)$ is given by choice function $C(C^*(\cdot))$. This means that for a given $X \subseteq A$ a choice must first be made with function $C^*(\cdot)$, and then with function $C(\cdot)$.

Definition

With respect to function $C^*(X)$, choice function $C(X)$ satisfies the condition:
of independence from excluded variants not belonging to function $C^*(\cdot)$ (I R condition), if

$$C(X) = C(X \setminus X') \quad \text{where} \quad X' \cap C^*(X) = \emptyset;$$

of representability like superposition (R S condition) if

$$C(X) = C(C^*(X)) \qquad \forall X \subseteq A;$$

of commutativity of superposition (S C condition) if

$$C(C^*(X)) = C^*(C(X)) \qquad \forall X \subseteq A;$$

of amplified commutativity (A S C condition) if

$$C(X) = C(C^*(X)) = C^*(C(X)) \qquad \forall X \subseteq A;$$

of inverse representability like superposition (I R S condition) if

$$C(X) = C^*(C(X)) \qquad \forall X \subseteq A.$$

The characteristic conditions we have now presented can be easily interpreted. For example, when Pareto's choice function is considered as function $C^*(\cdot)$, condition I R requires that the variants not belonging to the Pareto set do not influence the choice.

Some theorems establishing reciprocal relations among the newly introduced characteristic conditions are presented in the sequel.

Theorem 3

If $C^*(\cdot)$ satisfies condition O, then $C(\cdot)$ satisfies condition R S, with respect to $C^*(\cdot)$ if, and only if, function $C(\cdot)$ satisfies condition I R with respect to $C^*(\cdot)$.

Theorem 4

If $C^*(\cdot)$ satisfies condition C M, then $C(\cdot)$ satisfies condition I R S.

From theorem 3 and theorem 4 it follows:

Corollary

$C^*(\cdot)$ satisfies condition C M and 0. Then, function $C(\cdot)$ satisfies condition I R S with respect to $C^*(\cdot)$ if, and only if, $C(\cdot)$ satisfies condition I R with respect to $C^*(\cdot)$.

Theorem 5

Function $C^*(\cdot)$ satisfies condition H C and 0. Then, for any function $C(\cdot)$ such that $C(X) \subseteq C^*(X)$, conditions I R, R S, S C and A S C are equivalent.

It is well known that the domain $H \cap C \cap O$ [1] in the space of the choice functions is produced only by Pareto's functions. Then, from theorem 5, it follows that for any function $C(\cdot)$ isolating the Pareto set, conditions I R, R S, S C and A S C are equivalent. In particular, these functions are the tournament matrices

$$\left\{ C_s(\cdot), \quad s \in [1, \infty] \right\}.$$

We can demonstrate that choice function $C_{KR}(\cdot)$ satisfies condition I R and, due to theorem 5, it satisfies conditions R S, S C and A S C. Other functions

$$\left\{ C_s(\cdot), \quad s \in [1, s_0], \ s_0 < \infty \right\},$$

do not satisfy these conditions.

In the "multi-criterial" choice, a procedure is often used, for which the choice is made on the basis of the "proximity" of the variants to some "ideal" point.

Such models are studied in ref. [1]. Functions $C(\cdot)$ generated by this procedure do not satisfy conditions H, C or 0, but they can be described using the new characteristic conditions we have previously presented.

NOTES:

1: This theory deals with the results of using aggregation procedures of single evaluations in terms of reasonableness, that is, the properties of choice functions corresponding to the procedures at issue are considered.

REFERENCES

1. Aizerman M. A., Malishevski A. V.: 'General Theory of Best Variants Choice: Some Aspects' - IEEE Trans. Automat. Control, 26, 1030-1040 (1981)
2. Yemel'yanov S. V., Borisov V. I., Malevich A. A., Cherkashin A. M.: 'Models and Methods of Vector Optimization' - Tekhnicheskaya Kibernetika (Itogi nauki i tekhniki) (Control Engineering Results in Science and Technology) (in Russian) 5, 386-448 (1973)

# C-CALCULUS: AN OVERVIEW

E.R.Caianiello
Dipartimento di Fisica Teorica
e sue metodologie per le Scienze Applicate
Università di Salerno
84100 Salerno

## 1. A CALCULUS OF PARTITIONS

a) We propose to present some applications to Pattern Recognition of a "mathematical game" started some years ago by one of us[3] which was named C-calculus for reasons which will be reminded in the sequel. We wish to state forthwith that it is simpler in principle than ordinary arithmetics; various fields can be envisaged in which it might prove of use: e.g. manifold topology, integration theory, fuzzy sets (where it might provide a natural tool for numerical computation), measure theory in physics, data-base structures, neural models, etc. This, we hope, will be apparent to the reader; we must restrict ourselves here only to the specific field of interest in the present context. We shall endeavour to keep language and arguments as plain as the subject really is; recourse to abstract formalism is often a disguise more convenient to the author than to the reader. We begin therefore by remembering the game with which it all started. Take any integer positive numbers, and apply to them the rules of arithmetics, with the restrictions that only the <u>direct</u> operations, sum and multiplication, be allowed; the inverse ones, subtraction and division, forbidden; define furthermore the sum and the product of any two digits as follows

$$(1) \quad \begin{aligned} a + b &= \max(a,b) \\ a \times b &= \min(a,b) \end{aligned}$$

We may thus "multiply" any two such numbers, e.g. 736 and 491

```
  736x          491x
  491=          736=
  111           461
 736           331
434           471
47461         47461
```

We find that multiplication (and addition) thus defined are always



*E. R. Caianiello and M. A. Aizerman (eds.), Topics in the General Theory of Structures, 163–173.*
*© 1987 by D. Reidel Publishing Company.*

commutative for any such "numbers".

It would be an easy matter to demonstrate that, provided the "single digit" operations (1) are meaningful, one can operate in the same way on objects (subtraction and division being of course barred), such as vectors, matrices, etc., obtaining additivity and commutativity whenever they hold in arithmetics.

These "numbers" or "strings" of digits, with the operations(1), form clearly a commutative semi-ring. As in arithmetics, each "digit" plays two different roles: one intrinsic to it ("cardinality"), the other ("position") relative to the string in which it belongs. The next remark is that standard set theory treats only intrinsic properties of sets. If in (1) we interpret + as "union" $\vee$ and x as "intersection" $\wedge$, we can immediately transport all that was said thus far to "strings of sets", or "composite sets", "C-sets" for short.

Operating on C-sets as before, with $\vee$ and $\wedge$ in place of + and x in (1) (a,b denote now the "simple" sets of which C-sets are strings, as the digits in the former example), one has "C-calculus": a commutative semi-ring which permits, from some given C-sets, the generation of any number of other C-sets. Inverse operations are neither possible nor required in this context: only direct ones are permissible; one may perhaps see, though, advantages in being able to express in this way long lists of specifications in terms of a few basic ones.

An example of C-operation of special relevance for our present purpose is the following. Consider a segment S paritioned in segments $a_1, a_2, a_3, \ldots\ldots\ldots, a_k$; this partition $A = a_1 a_2 \ldots\ldots\ldots a_k$

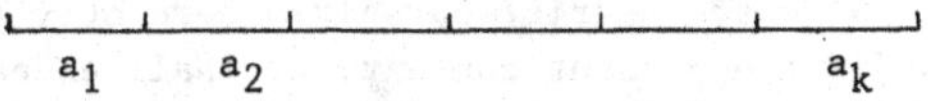

$a_1 \quad a_2 \qquad\qquad a_k$

Consider now the same segment partitioned in a different way $B = b_1 b_2 \ldots\ldots\ldots b_\ell$

$b_1 \quad b_2 \qquad\qquad b_\ell$

Consider now A and B as C-sets: the elements of each partition or string, are "simple" sets; C-multiplication of A and B gives

$$A \times B = B \times A = a_1 a_2 \ldots a_k \times b_1 b_2 \ldots b_\ell = C = c_1 c_2 \ldots c_p$$

and it can be verified immediately that the simple sets of the product are obtained, in order, by joining on the segments the terminal points of both partitions A and B.

The C-product of two partitions gives thus the refinement of one by the other: C-calculus is the natural way of composing partitions, or coverings. In fact, the same property holds true in any number of dimensions, [1] . This is the key property of C-calculus as regards its application to Pattern Recognition.

b) Our interest in an approach of this type to Pattern Recognition originates from the instinctive feeling of the physicist when confronted with problems for which a vast number of approaches are proposed, some indeed of remarkable ingenuity and power, but none of general (at least in some sense) applicability: is there some philosophy, or method, that may be applied to all problems of this sort, even if, of course, with less abundance of results than ad hoc techniques will undoubtedly provide? Does one need a language for thinning, one for shrinking, one for counting, one for studying textures, one for retrieving objects against a background, and so on? Or may we, perhaps, let "patterns" speak for themselves, changing the pattern itself into something algebraic or numeric, out of which several, if not all, questions may be answered through essentially a single basic algorithm? This is, of course, a "träumerei; but the search for "laws", rather than "rules", is a professional deformation for which a physicist need not apologise, although he had better be - as we certainly are - duly apologetic about results achieved.

Having expressed (not certainly justified) our motivation, we shall substantiate it with a typical instance. Granted a priori that a major crime of Pattern Recognition is the preliminary reduction of a (say) 2-dimensional image into pixels, and that we must so proceed because we are much less bright than a fly or a frog, we find that a rather peculiar situation then arises. Parcelling a picture into pixels (with tones of grey, or colour) is, logically, a parallel process, out of which we can gather more information, the finer the grid whose windows generate "homogenized pixels" (from each pixel only averages are taken). Suppose now that the same grid is rigidly shifted, over the picture, by a fraction of its window size; we may proceed as before, and obtain some other amount of parallel information. The first question arises: can we use both pieces of information, the one from the first and the one from the second grid partitioning, to get better, more detailed information on the picture? Since we are taking only averages from each pixel, each time, the answer is no (unless, of course, we perform some mathematical acrobatics): one of the two readings has to be thrown away.

It would be nicer, one might feel, if there were a way of performing readings from the grid such as to permit the combination in a natural way of the readings of both grids to obtain more refined information on the picture (as might have been gathered by using a finer grid to begin with). If one can handle this situation, conceivably it may then be possible to use several times (serially) a single coarser grid, out of which is read the (parallel) information obtained by shifting the whole grid by one step, and so on. The use we intend to make of C-calculus is aimed at answering just this question. The "reading" from a grid (of a suitable sort) becomes per se a C-set; two C-sets from different positions of the grid can be C-multiplied: this will give finer information, and so on.

c) Under "suitable" circumstances (to be defined explicitly in the sequel) this procedure can be carried through to the extreme limit of perfect reconstruction of the original picture (as digitized at the finest possible level: e.g. with a $2^{10} \times 2^{10}$ grid for the original, it may be reconstructed by covering it stepwise with, say, a $2^3 \times 2^3$ grid). During this process, many things which one does with specific techniques such as contour extraction, contrast enhancement, feature extraction etc., can be performed by interpolating in it steps which "answer" questions of this sort and become part of the algorithm. But the application of C-calculus will often fail; the original image may not be thus reconstructed. There is an element to be considered, which was before ignored with the adjective "suitable": the size of the window. It is a feature of our approach that the critical size, below which total reconstruction of the picture is impossible, is determined by the structure of the picture itself, and is not a matter of guesswork or trial and error.

One can arrange readings, and ways of analyzing them, from grids having sizes appropriate to constitute filters that see some wanted features and are blind to others. Typically, consider a saucer on a chessboard: things can be arranged so as to see only the saucer or only the chessboard (with a hole); or a specific component of a texture, ignoring all others, or to suppress some noise etc. Such filtering does not smear out or enhance; it gives at worst an indented contour to the saucer, as is natural when working with grids.

A study of this filtering process from the point of view of rigorous mathematics has not been undertaken yet, again from the physicist's point of view that such studies are always possible on no matter what subject, but that it is preferable first to test whether the subject is worth the effort.

At this stage a wealth of tools become available, about which it is better to keep a critical rather than an enthused view. Most times the practical problems at hand require only partial answers for their solution, like distinguishing between given printed characters.

## 2. CONVERGENCE AND FILTERING

Our main tool in the application of C-calculus to Pattern Recognition will be, as was said at the end of section 1a), the C-multiplication of two partitions. A and B of an N-dimensional domain

$$A = a_1 \ldots\ldots\ldots a_k$$
$$B = b_1 \ldots\ldots\ldots b_\ell \qquad \text{which yields}$$

$$C = A \times B = B \times A = c_1 \ldots\ldots\ldots c_p$$

i.e. is the refinement of A by B, or viceversa, with elements $c_1 c_2 \ldots c_p$
The same approach can be used regardless of the number of dimensions of the picture to be studied. We shall consider, for the sake of simplicity and for specific relevance to P.R., only one- and two-dimensional patterns (e.g. graphs and pictures); we shall also consider only one additional dimension, which may for example denote levels of greyness, discretized or not (with colours, one may add as many dimensions as distinct ones are considered, etc.).

We start with one-dimensional patterns, as in this case the procedure can be visualized immediately. We restrict our attention to "grids" i.e. to partitions of the abscissa x into segments, for simplicity, of equal length, the "windows".

Thus, consider the graph of fig.1, where the ordinate y denotes greyness (or intensity of sound, or local pitch...). Discretize now the x coordinate with a grid G, of window w; change the graph into a sequence of rectangles, by substituting the portion of graph corresponding to a given window $w_h$ with the rectangle which projects upon w into $w_h$ and upon y into the segment having as upper and lower extrema the maximum $M_h$ and the minimum $m_h$ reached by the graph within $w_h$ (no matter where, or how many times.). We change thus the graph into a string of rectangles.

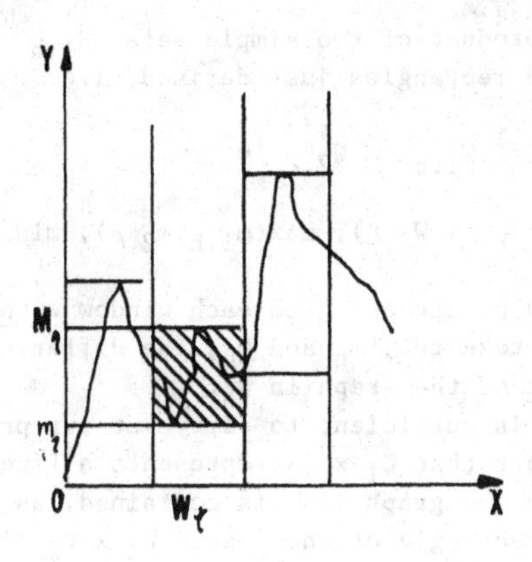


Fig.1 A C-set: a partition of a graph in a string of quadruples

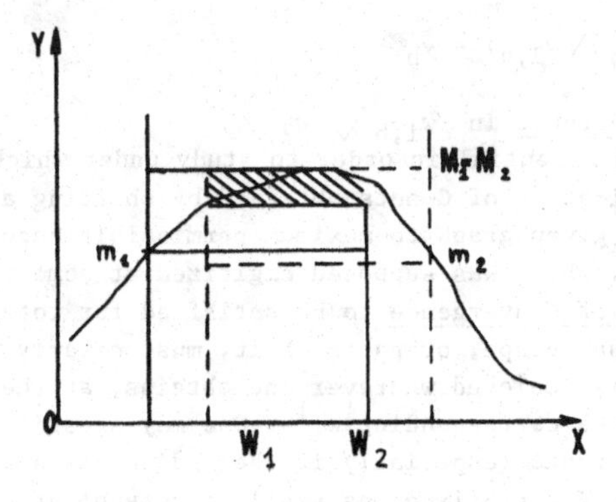


Fig.2 A product of the elements of two C-sets

The ordered sequel of all these rectangles is the C-set determined from the graph by partitioning the x-axis with the given grid. We may now proceed as before, after shifting the grid by a step $1 < w$. Denoting with indices 1 and 2 the two C-sets thus obtained, we have, with an obvious notation

$$C_1 = R_{1,1},\dots\dots\dots,R_{1,k}$$

$$C_2 = R_{2,1},\dots\dots\dots,R_{2,k}$$

$$R_{i,h} = (w_{i,h};m_{i,h},M_{i,h})$$

We may now define the product of two simple sets $R_{1,h}$ and $R_{2,\ell}$ as the intersection of the rectangles just defined, i.e.

$$R_{1,h} \times R_{2,\ell} = \begin{cases} \phi \text{ iff } W_{1,h} \wedge W_{2,\ell} = \phi \\ (W_{1,h} \wedge W_{2,\ell}),\ \max(m_{1,h},m_{2,\ell}),\ \min(M_{1,h},\ M_{2,\ell}) \end{cases}$$

In other words; instead of attaching to each window w <u>one</u> value, say the average of y in it, we take <u>two</u>, $m_h$ and $M_h$; the difference $M_h - m_h = \Delta_h$ is known as the <u>dynamic</u> of the graph in $w_h$.

This modification is sufficient to carry out our proposed programme, because now it is evident that $C_1 \times C_2$ represents a finer partition of some strip within which the graph-line is contained, as is shown in fig.2.

Consider now any rectangle of the C-set $C_1 \times C_2$; its base is $w_{1,h} \wedge w_{2,h}$, the height is

$$y_h = \min(M_{1,h},\ M_{2,h}) - \max(m_{1,h},\ m_{2,h}) \quad .$$

It is thus evident that the dynamic of the graph in it is:

$$\Delta(w_{1,h} \wedge w_{2,h}) \leqslant y_h \qquad \text{which reduces}$$

$$\Delta(w_{1,h} \wedge w_{2,h}) = y_h$$

if the graph is <u>monotonic</u> in $w_{1,h} \vee w_{2,h}$ .

This remark is essential in order to study under which conditions iterated C-multiplication of C-sets obtained by shifting a given grid will reproduce the given graph to maximum permissible accuracy (that of the original graph, which was supposed digitized at some finer level).

The <u>criterion of convergence</u> to be satisfied for total reconstruction of the original graph, or parts of it, must clearly be the following: convergence is achieved wherever one obtains, at the $h^{th}$ iteration, $m_{i,h} = M_{i,h}$ over the corresponding $w_i$'s. One may substitute this criterion by the weaker one (especially if the ordinate is not discretized) that $M_{i,h} - m_{i,h} < \varepsilon$ , prefixed, as small as convenient.

The formalization of this procedure is straightforward; the interested reader may reproduce it by himself, or refer to some previous paper [2 ].

For equally spaced grids (it would be only a matter of convenience to relax or change this condition at any desired step of the procedure), there is a very simple formula which determines whether overall convergence is guaranteed: this will be the case if, and only if:

$$W \leqslant \frac{D}{2} + 1 \qquad (2)$$

where D denotes the smallest distance between a minimum and a maximum of the graph. The proof is given in [2] . The same formula applies also in two (or more) dimensions if we now read w to mean the side of the square window, and D the euclidean distance in the plane between such extrema.

b) We can now change our viewpoint. Instead of shifting the grid, obtaining and multiplying the ensuing C-sets, etc., we ask what this procedure will finally yield at a given, fixed point on the x-axis. The answer is that, regardless of the order in which these operations are performed, the final values which are associated with any given point are the highest minimum and the lowest maximum that are seen when the window w moves on an interval of length 2w - 1, centered at that point. The discussion of this point, which is related of course to (2), is in [2] .Of special interest is the case in which (2) is violated. Our procedure will not reconstruct then the original pattern, but produce a new pattern, which suppresses all those details of the original one which could be retrieved only by respecting (2). In other words, the procedure will act now as a filter (C-filter [5] ). A trivial, intuitive example will convince us of this fact. Suppose that we wish to study in this way the above mentioned chessboard. Operating with a square grid whose window is smaller than the case of the chessboard, we will readily reconstruct the chessboard. If, however, the window is larger than the case, no matter how we move the grid we shall always find m = 0, M = 1 (say) in any window: convergence is impossible. It is then a trivial matter to arrange things so that in the first case our procedure reconstructs the chessboard, in the second it yields a total blank: the chessboard is filtered away. If we have a saucer on the chessboard, we shall we able to retrieve only the saucer, obliterating the chessboard background; likewise, one can proceed with textures [4] : it is possible to see an object ignoring a textural background, or viceversa, to extract only some relevant textural elements. Many variations are possible on this theme [7] . One can keep, thus, a window which satisfies (2) but only accept dynamics within given thresholds ; or play both with window size and threshold.

## 3. EXAMPLES

A few examples will suffice here.

### 3.1 Contour Extraction

This is a relevant step in every problem of P.R., with a wide literature and a large number of techniques available: gradients and thresholds, thinning, joining or separating of broken or interpenetrating pieces, template matching, etc. [13,11,8]. The figures show some results with our technique.

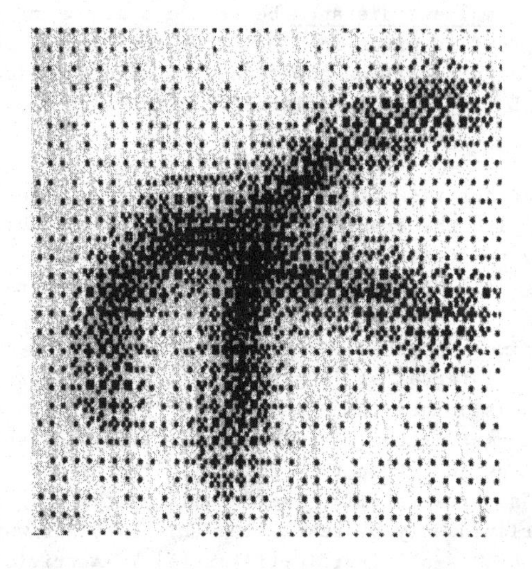

Fig.3 Human chromosome

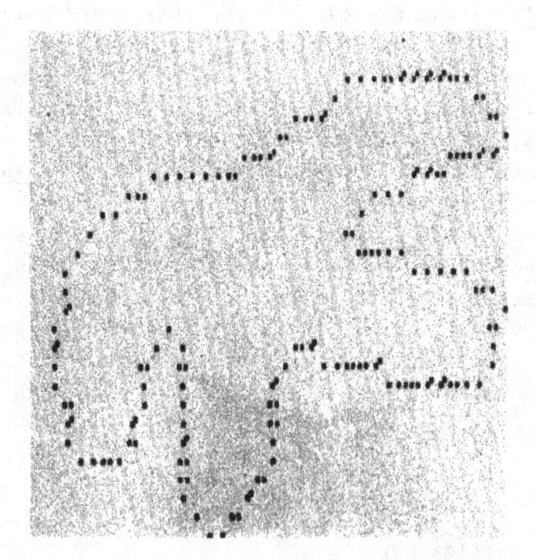

Fig.4 Contour extraction of a human chromosome

### 3.2 Textures

The analysis of textures, both <u>per se</u> and as backgrounds, is almost a science within a science; [9,12,6]. Our views on this subject do not belong to this short summary of our work; more than in other topics of P.R. we must remind ourselves that "percepta" are "phenomena", in the sense of Kant, quantum mechanics or the Vedas, to which the "noumena" and the "observer" equally concur. We wish only to report here, in conclusion, that C-calculus has proved especially "natural" in this context, leading to ready and elementary classification, analysis and discrimination.

We refer the reader to previous works [10] ; some examples are reported in this. It is the "chessboard and saucer" game mentioned earlier which is easily implemented through C-calculus into algorithmic simulations which may be rather close in principle to the actual operation of neural tissues.

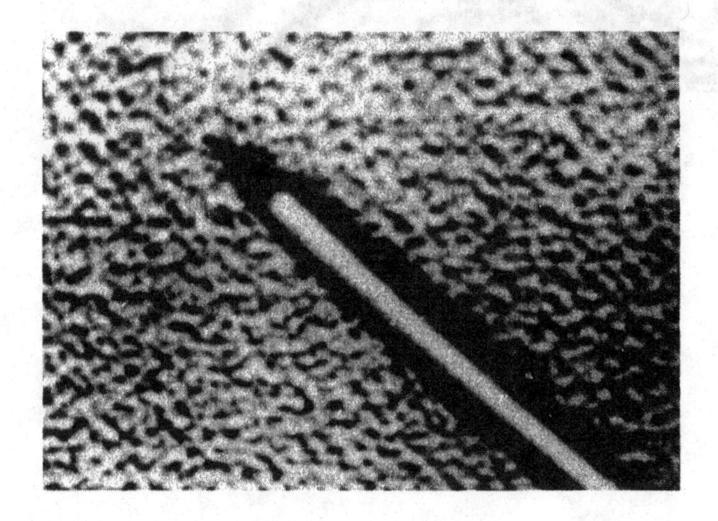

Fig.5 A pencil on a natural texture

Fig.6 The application of our filtering method

REFERENCES


1. A.Apostolico, E.R.Caianiello, E.Fischetti, S.Vitulano: 'C-calculus: an elementary approach to some problems in Pattern Recognition', Pattern Recognition, 5 (1978).
2. A.Apostolico, S.Vitulano: 'An image transform', in Informatik-Fach berichte, 8, Springer Verlag (1977).
3. E.R.Caianiello: 'A calculus for hierarchical systems', Proc.First Int.Joint Conf.on Pattern Recognition, Washington (1973).
4. E.R.Caianiello, A.Gisolfi, S.Vitulano: 'A technique for texture analysis using C-calculus', Signal Processing, 1 (1979).
5. E.R.Caianiello, A.Gisolfi, S.Vitulano: 'A method of filtering bio-medical specimens', Proc.The Int.Conf.of Cybern.on Society, Tokyo (1978).
6. I.Dinstein, R.M.Haralick: 'Textural features for image classification', Proc.IEEE trans.of System, Man and Cybernetics, 6, 73.
7. M.Galloway: 'Texture analysis using grey level run lengths',Comput. Graphics Image Process, 2 (1978).
8. F.V.German, M.B.Clowes: 'Finding picture edges through collinearity of feature point', IEEE trans.Comput., 24.
9. J.J.Gibson: The perception of the visual world, Houghton (1950).
10. A.Gisolfi, M.Mlodkowsli, S.Vitulano: 'A method for classifying and filtering textures', Progress in Cybernetics and System Research, Hemisphere Pub., Washington (1980).
11. E.Hall, W.Frei: 'Invariant features for quantitative scene analysis', Univ. Southern California (1976).
12. W.Köhler: La psicologia della forma, Feltrinelli (1975).
13. L.G.Roberts: Machine perception of three dimensional solids, M.I.T. Press (1965).

# ON SOME ANALYTIC ASPECTS OF C-CALCULUS

Eduardo R.Caianiello (1)
Aldo G.S.Ventre (2)

(1) Dipartimento di Fisica Teorica
e sue Metodologie per le Scienze Applicate
Università di Salerno
84100 Salerno

(2) Istituto di Matematica
della Facoltà di Architettura
Università di Napoli
Via Monteoliveto 3
80134 Napoli

INTRODUCTION. In the description and analysis of various systems there are some situations in which properties or readings cannot be considered as sharp. For instance, this happens in hierarchical classifying, in describing physical phenomena for which the measurements have to be considered valid in ranges rather than in single points and, generally, when some data of the problems are given with tolerances.

Often precision is both unrealistic - and for over-complex systems is not attainable - and not to be hoped for because too many constraints result in a problem without solutions [8]. A technique, called composite-set calculus, or for short, C-calculus, was introduced in [2], in order to deal with problems whose solutions have several degrees of acceptance depending on the added quantities of information.

Among the areas of applications of C-calculus we mention the analysis of hierarchical systems [2], a number of fields in pattern recognition (see, for instance [1], [4]) and mechanical design [5]. C-calculus assumes that the k-th state of a knowledge analysis system is generated by the previous ones in such a way that the reading of some characters is "improved".

Roughly speaking, improvement means suitable maximization or minimization of parameters, as we shall define in the next section.

## 1. A MATHEMATICAL FRAMEWORK FOR C-CALCULUS

Let X and S be non-empty sets. If a non-empty subset E of S is associated with each element x in X, then a function f from X to the set of



E. R. Caianiello and M. A. Aizerman (eds.), Topics in the General Theory of Structures, 175–181.
© 1987 by D. Reidel Publishing Company.

subsets of S is defined. Often f is also called a correspondence from X to S, and one writes $E = f(x)$.

We first suppose X to be a topological space and S a totally ordered set. Let f be a correspondence from the open sets of X to S. Let $(A_i) = (A_1,\ldots,A_m)$ and $(B_j) = (B_1,\ldots,B_n)$ be partitions of X into open non-empty sets. The non-empty entries of the matrix $(A_i^1 \cap B_j^1)_{i,j}$ form a new partition of X. For $A_i \cap B_j \neq \emptyset$, it is

$$\left.\begin{matrix}\inf f(A_i)\\ \inf f(B_j)\end{matrix}\right\} \leq \inf f(A_i \cap B_j) \leq \sup f(A_i \cap B_j) \leq \left\{\begin{matrix}\sup f(A_i)\\ \sup f(B_j)\end{matrix}\right.$$

The values

$$\bar{m}_{ij} = \max(\inf f(A_i),\ \inf f(B_j))$$

$$\bar{M}_{ij} = \min(\sup f(A_i),\ \sup f(B_j))$$

are readings or descriptions related to the intersections $A_i \cap B_j \neq \emptyset$. We take them to be actual "improvements" of the description given by f. Such a point of view does not lead at present to average operations. If a new partition into non-empty open sets is given for X, $(C_h) = (C_1,\ldots C_p)$, then a finer partition of X is obtained, with non-empty elements $(A_i \cap B_j \cap C_h)_{i,j,h}$, and so on. Then for every subset $A_i$ of $(A_i)$ a decreasing sequence of sets

$$(1) \quad A_i \supset A_i' \supset \ldots \supset A_i^{(n)} \supset \ldots$$

is defined, by means of the repeated intersections. Let us consider just the non-empty elements of the sequence (1).

Therefore, if we set

$$m^1 = \inf f(A_i), \quad m^2 = \bar{m}_{ij}, \quad m^3 = \max(m^2, \inf f(C_h)),\ldots$$

$$M^1 = \sup f(A_i), \quad M^2 = \bar{M}_{ij}, \quad M^3 = \min(M^2, \sup f(C_h)),\ldots$$

then it is

$$m^1 \leq m^2 \leq \ldots \leq m^k \leq \ldots \leq M^k \leq M^{k-1} \leq \ldots \leq M^1$$

If

$$\sup_k \left\{m^k\right\} = \inf_k \left\{M^k\right\}$$

then the process is said to be convergent. In this case the characters described by f are precisely known for the last non-empty set A of (1), if there is any, or for the (non-empty) set $A = \lim_n A_i^{(n)}$. Otherwise indeterminacy is not suppressed in A.

## 2. A MEASURE-THEORY MODEL

We shall now consider the special case when the correspondence f takes values in closed intervals of the real unit interval $I = [0,1]$ . Some measure-theoretical hypotheses are supplied in order to focus attention both on some convergence and non-convergence cases, that occur, in particular, in pattern recognition. We shall find also that non-convergence cases are equivalent to filtering procedures. Some terminology of interval analysis is required for such an approach and can be found in the book by Moore [7] .

Let $(X,\mathcal{A},\mu)$ denote a measure space with $\mathcal{A}$ a $\sigma$-algebra of subsets of X and $\mu$ a totally finite positive measure. (Measure theoretical terms can be found in the book by Hewitt and Stromberg [6] ). Consider now a measurable dissection of X, $(X^\circ) = (X_1^\circ,\ldots,X_p^\circ)$, i.e. a finite, pairwise disjoint family of elements of $\mathcal{A}$ such that their union coincides with X. Let $g : X \longrightarrow I$ be a function,

$$m_i^\circ = \inf_{X_i^\circ} g(x), \qquad M_i^\circ = \sup_{X_i^\circ} g(x),$$

and consider the functions

$$m^\circ(x) = \sum\nolimits_i m_i^\circ c_i(x)$$

$$M^\circ(x) = \sum\nolimits_i M_i^\circ c_i(x)$$

where $c_i$ is the characteristic function of $X_i^\circ$. The triple $((X^\circ),m^\circ,M^\circ)$ is, by definition, a C-set (induced by g) in X.

Let now $(X') = (X_1',\ldots X_n')$ be another C-set in X, and define analogously the functions $m'(x)$ and $M'(x)$. The set of all non-empty entries of the matrix $(X_i^\circ \cap X_j')$ , $i=1,..,p$ and $j=1,..,n$, is a measurable dissection of X. This partition is a refinement of each of the previous partitions. Let us define

$$(2) \quad m_{ij}^1 = \max(m_i^\circ,m_j'), \quad M_{ij}' = \min(M_i^\circ,M_j'),$$

and the simple functions

$$m^1(x) = \sum\nolimits_{ij} m_{ij}^1 c_{ij}(x)$$

$$m^2(x) = \sum\nolimits_{ij} M_{ij}^1 c_{ij}(x),$$

where $c_{ij}$ is the characteristic function of $X_i^\circ \cap X_j^\circ \neq \emptyset$. The binary operation * in the set of all C-sets in X, defined by

$$((X^\circ),m^\circ,M^\circ) * ((X'),m',M') = ((X^1),m^1,M^1)$$

is called C-multiplication and the right-hand side is the C-product.

The C-products of the C-sets induced by the function g are not in general C-sets induced by g. However, the meaning of * can be enlarged as is shown in [ 3 ], by applying repeatedly the rules (2) in order to define the C-multiplication between a C-set and a C-product, between C-products and so on. Indeed the result is even a triple $((\bar{X},a,b,))$ with $(\bar{X})$ a measurable dissection of X, $a(x) = \sum a_r c_r(x)$, $b(x) = \sum_r b_r c_r(x)$, where $0 \leq a_r \leq b_r \leq 1$. Thus a subset Cg, is defined in the set of all C-sets in X: it is the set of all finite C-products of C-sets induced by g. The structure (Cg,*), is a commutative monoid, whose identity is $(X,\underline{0},\underline{1})$, the C-set having the one element dissection where $\underline{0}$ and $\underline{1}$ are the constant functions taking values 0 and 1 in X, respectively. Consider now the sequence $\{C_k\}$ of C-products defined by :

$$
\begin{aligned}
(3)\quad & ((X^\circ),m^\circ,M^\circ),\\
& ((X^1),m^1,M^1) = ((X^\circ),m^\circ,M^\circ) * ((X'),m',M'),\\
& ((X^2),m^2,M^2) = ((X^1),m^1,M^1) * ((X''),m'',M''),\\
& \dots\dots\dots\dots\\
& ((X^k),m^k,M^k) = ((X^{k-1},m^{k-1},M^{k-1}) * ((X^{(k)},m^{(k)},M^{(k)})\\
& \dots\dots\dots\dots
\end{aligned}
$$

It is easy to see that, if $X_s^k$ is the general non-empty element of $(X^k)$, then for every $x \in X_s^k$,

$$m^\circ(x) \leq m^1(x) \leq \dots \leq m^k(x) \leq g(x) \leq M^k(x) \leq \dots \leq M^1(x) \leq M^\circ(x),$$

where $m^k(x) = \sum_s m_s^k\, c_s{}^k$ and $M^k(x) = \sum_s M_s^k\, c_s^k(x)$.

As we sketched in the introduction, from transition equation (3), it makes sense to talk about an "improvement" carried out by k-th state with respect to the previous one. Indeed, the differences $M^k(x)-m^k(x)$ measure step by step just these improvements. We stress that the cases when the sequence $M^k(x)-m^k(x)$ takes a constant non-zero value, from a certain $\bar{k}$ on, are also worth analyzing both from a theoretical and a practical point of view. This will be dealt with in the sequel.

Let $(X^k)$ be a measurable dissection of X and x an element of X. There is a unique element $X_s^k$ of $(X^k)$ which contains x, and, for every $x \in X$, a sequence

$$I_{s0}^\circ \supseteq I_{s1}^1 \supseteq \dots \supseteq I_{sk}^k = [m_{sk}^k, M_{sk}^k] \supseteq \quad \dots$$

of intervals is determined. If the sequence $\{I_{sk}^k\}$ converges to a real number c, then the sequence $\{C_k\}$, $C_k = ((X^k),m^k,M^k))$, is convergent, i.e. represents a convergent process(see introduction). If the sequence $\{I_{sk}^k\}$ converges to a non-degenerate interval $A = [a,b]$, then $\{C_k\}$ is non-convergent in x. Exactly one of the two cases occurs for every $x \in X$.

(The sequence $\{C_k\}$ is, by definition, convergent (non-convergent) in a subset A of X if it is convergent (non-convergent) in every element of A). Let now $X_s^k$ be the general element of $(X^k)$ and put

$$L^k(g) = \sum_s \mu(X_s^k) m_s^k, \quad U^k(g) = \sum_s \mu(X_s^k) M_s^k.$$

On the other hand, the lower and upper Darboux sums of g, with respect $(X^k)$, are

$$\bar{L}^k(g) = \sum_s \mu(X_s^k)\, \bar{m}_s^k, \quad \bar{U}^k(g) = \sum_s \mu(X_s^k)\, \bar{M}_s^k,$$

where $\bar{m}_s^k = \inf_{X_s^k} g$ and $\bar{M}_s^k = \sup_{X_s^k} g$

The following inequalities hold

$$(4)\quad \bar{U}^k(g)-\bar{L}^k(g) \leqslant U^k(g)-L^k(g) \leqslant \sum_s \mu(X_s^k)(M_s^k - m_s^k) \leqslant \mu(X).$$

We are now able to state a first criterion of convergence of $\{C_k\}$.

Proposition 1: If for every $\varepsilon > 0$ there is an integer k' such that $k>k'$ implies

$$\sum_s \mu(X_s^k)(M_s^k - m_s^k) < \varepsilon,$$

then $\{C_k\}$ is convergent almost everywhere in X.

Let us consider some cases of convergent sequences $\{C_k\}$ in X.

a) If

$$(5)\quad \sup_s \mu(X_s^k) \longrightarrow 0 \qquad \text{as} \quad k \longrightarrow \infty$$

and, for a positive real number r, $\sum_s (M_s^k - m_s^k) \leqslant r$, for every k, then $\{C_k\}$ is convergent in X.

Let s(k) denote the number of the addends in the sum in (4), and

$$d_k = \sup_s (M_s^k - m_s^k) \quad \text{and} \quad e_k = \sup_s \mu(X_s^k) .$$

b) If

$$(6)\quad 0 < \frac{e_k}{e_{k-1}} \leqslant a < 1$$

and $$0 < \frac{d_k}{d_{k-1}} \leqslant b < 1, \text{ then}$$

$$\sum_s \mu(X_s^k)(M_s^k - m_s^k) \leqslant e_o d_o s(k) a^k b^k$$

If one of the products $s(k)a^k$, $s(k)b^k$ is less than a constant B, for every k, then the sequence is convergent almost always in X. This happens for instance, when every element of the dissection is halved at each step of the procedure and the number of the elemnts in $(X^k)$ is $p2^k$.

c) If (6) and $d_{k-1} = d_k$ hold, for every k, then

$$\sum_s \mu(X_s^k) d_s^k \leqslant d_o \mu(X) .$$

Then, if the further condition $d_o = M_s^k - m_s^k$, holds, for every k, then $\{C_k\}$ is non-convergent almost everywhere in X.

d) If $0 < c \leqslant \bar{d}_k = \inf(M_s^k - m_s^k)$, for every k, then $\{C_k\}$ is non-convergent almost everywhere in X.

An interesting property is that there are some sequences $\{C_k\}$ such that X can be partitioned into a subset E, where $\{C_k\}$ is non-convergent, and X-E, where it is convergent. Then it can sometimes be advantageous to construct sequences $\{C_k\}$ which are convergent or not in a fixed part of X, according to particular purposes. When, for instance, in every point x of a subset E of X the sequende $\{I_{sk}^k\}$ does not converge to a single point, then the exact description of x is not achieved. In other words, a procedure of repeated non-convergent C-products $C_k$ in a subset E of X acts as a filter on E. An example will be discussed in the next section.

The interval $I_{sk}^k$ is a function of $x \in X$ and shows the lack of precision in measurement, or reading, related to X. The global imprecision carried by the C-set $C_k$ - i.e. the amount of the single contributions to imprecision of the reading $C_k$, given by the widths of the intervals $I_{sk}^k$ on every $x \in X$ - is measured by

$$\sum \mu(X_s^k)\ (M_s^k - m_s^k) \quad .$$

## 4. SOME APPLICATIONS TO PATTERN RECOGNITION

Let us consider rectangular dissections of X (see [ 1 ] ).

Let a digitized picture in the plane xy be represented by a matrix $A = (a_{ij})$, whose entries yield the grey levels $g = g(x,y)$. Furthermore a window is provided such that the maximum M and the minimum m of the grey levels are observed over the rectangular area w, viewed through the window. If the picture is scanned, so that the matrix A is completely covered, all addends of the sum in (4), for k = 1, are known.

The picture is approximated by a grid superimposed on it, which shows, for each area w, the bounds of the grey levels. Translate now the grid so that a different partition of the picture is read. The C-product of the two readings takes into account all pieces of information of both of them: a refinement of the distribution of the grey levels in the picture is achieved or not depending on the width of the window, the width of the translation of the grid which defines the sequence $\{\mu_k\}$, and the differences $M_s^k - m_s^k$ between the grey levels.

Let us mention the typical filtering procedure of 'the saucer on the chessboard'. If the region viewed through the window includes strictly the single square, the sequence $\{C_k\}$ induced by the translations of the grid, acts as a filter because the values $M_s^k$ and $m_s^k$ over each region $X_s^k$ remain constant - say 1 and 0, respectively - for every k and s. (Case 3 or 4 apply). Thus only the saucer can be seen because the sequence of intervals $I_s^k$ is the sequence of real unit intervals $[0,1]$ ,for every x out of the saucer and on the chessboard.

REFERENCES


1. A.Apostolico, E.R.Caianiello, E.Fischetti, S.Vitulano: 'C-calculus: an elementary approach to some problems in pattern recognition', Pattern Recognition, 10, 375-387, (1978).
2. E.R.Caianiello: 'A calculus for hierarchical systems', Proc.Ist. Int. Congress on Pattern Recognition, Washington D.C. (1973).
3. E.R.Caianiello, A.G.S.Ventre: 'A model for C-calculus', in print.
4. E.R.Caianiello, Lu Huimin: 'A new approach to some problems of cell motion analysis based on C-calculus', see article in this volume.
5. A.Donnarumma: 'About the use of C-calculus in methodical design', Workshop Design-Konstruktion, 7, Zürich (1981).
6. E.Hewitt, K.Stromberg: Real and abstract analysis , Springer-Verlag Berlin, Heidelberg, New York (1965).
7. R.E.Moore: Interval Analysis , Prentice Hall, Englewood, Cliffs, N.J., (1966).
8. C.V.Negoita: Fuzzy Systems, Abacus Press, Tunbridge Wells, Kent (1981).

# A NEW METHOD BASED ON C-CALCULUS FOR SOME PROBLEMS OF CELL MOVEMENT ANALYSIS

Eduardo R. Caianiello and Lu Huimin*
*On leave from the Institute of Biophysics of Academia Sinica
Peking, People's Republic of China

INTRODUCTION. In the field of cellular biology cell movement has been up to now one of the less investigated areas. Cell movement includes cytoplasm and cytoplasmatic granules movement (e. g. the movement of granules in the pancreatic cell [1]); the movement of cilia and flagella of the cell; cell body movement (like leukocytes' amoeboid movement); cellular reproduction and so on.

Cell movement analysis basically consists of surveying position changes and alterations in dimension, shape and trama of objects, and of their monitoring.

Quantitative information about cell movement and its morphological changes is of particular interest to biologists. For example, in the study of cell morphology it is essential to understand the cell proliferation mechanism, which can have important effects on our knowledge of some kinds of cancer [2].

The knowledge of cell-reproduction methods is another matter of considerable importance; in this field, attention has always been concentrated on cell division, especially on mitosis, whereas other possible reproductive mechanisms have been neglected. S. Bei et Al. have proposed a new hypothesis for cell reproduction [3, 4].

They have suggested that cellular re-formation is another manner of cell reproduction, characterized by a cellular self-organization and self-assembling process.

Usually, experiments on living cells are carried out <u>in vitro</u>, and data are expressed in the form of a sequence of regularly spaced photograms. The quantity of data resulting from a single experiment is enormous; for example, for a single experiment, a non-automatic analysis required several weeks of human work, even for relatively simple [5] analyses. Furthermore, the repetitive and tedious nature of the work necessary for these analyses easily makes the procedure subject to human error, so that the use of a computer is very desiderable.

Here we intend to discuss mainly two problems: the processing of images and the detection of movement, both critical problems in cell movement analysis.

Our main concern is to try to use essentially one fundamental algorithm in order to solve many, if not all, problems of image processing



*E. R. Caianiello and M. A. Aizerman (eds.), Topics in the General Theory of Structures, 183–197.*
*© 1987 by D. Reidel Publishing Company.*

that arise in cell movement analysis.

Many different approaches to the various problems of image processing have been proposed so far (for filtering, extraction of contours, refinement, contraction and dilation, texture analysis, etc.), some really of remarkable ingenuity and power, but none of general applicability.

The present work is a preliminary step in this direction.

## THE C-CALCULUS

We begin by reminding the reader of some fundamental aspects of C-calculus.

C-sets ("composite" sets, hence the name C-calculus) were originally proposed in connection with the study of complex hierarchical systems [6] in order to establish a formalism which might allow a "natural" treatment of information concerning these systems. The main idea of C-calculus is based on the remark that in every numerical system the value of a figure depends both on its cardinality and its position. It is defined by the following formal rules:
Given the set of positive integers, operate on them with formal arithmetical rules, with the limitation that only direct operations, i. e. addition and multiplication, are allowed, whereas inverse operations, i. e. subtraction and division, are forbidden.

Furthermore, define the sum and the product of two digits as follows:

$$a \oplus b = \max(a, b) \qquad a \otimes b = \min(a, b) \qquad (1)$$

These operations have the commutative property. "Digits" can be substituted with "sets" (simple sets) and the operations in (1) with union and intersection. Under the operations (1), "sequences" or string of digits or simple sets become C-sets. These form commutative semirings.

C-sets can be obtained from a "digitized" image. In general, a "digitized" image appears like an n x m matrix A:

$$A = [a_{i,j}]$$

The elements of A represent the grey levels obtained by averaging over small rectangular areas of the figure.

A C-set for this matrix can be obtained by using a reading instrument with a window of amplitude W, which yields only the max and min values of grey in the area visible through the window.

With this reader we perform a scan of the image, moving the window successively to the next position (for example, consecutive horizontal strips). This scanning is equivalent to the division of the matrix A into submatrices of area W, each containing the max and min value of grey within the area W. This is equivalent to observing A through a grid.

All such submatrices are ordered; each contains a quadruple of values: a couple gives the position of the submatrix in the A matrix, the other couple the corresponding max and min grey-values.

Suppose that the grid has zero "initial phase"; the ordering of all these "quadruples" into a string gives set $C_0$. If we now translate rigidly the grid (this means assigning to the reader an initial phase different from 0), we obtain a second set $C_1$.

Multiplying $C_0$ by $C_1$, in accordance with the C-calculus rules, the result will give a finer partition of the original matrix and also a more precise description of the image it contains. Performing a certain number of these products among opportunely selected C-sets, we may expect to rebuild eventually an image with a degree of accuracy equal to that of the original.

The condition for the convergence of the above mentioned procedure in each point of the matrix is

$$W \leqslant 1 + D/2 \qquad (2)$$

where D is the length of the smallest interval in which the signal is monotonic [7]. Since C-calculus is commutative, C-products do not depend on the manner in which the scanning is performed: whether by lines, for example row by row, or column by column, or otherwise.

THE C-FILTER

In the processing of the image the distinction between object and background is an important and difficult problem. In some cases, the C-calculus can be used to filter on objects wanted by eliminating the background. This is the purpose of the C-filter. It has the advantage that the elimination of the points belonging to the background does not distort objects in any way.

The principle at the basis of this filtering method can be explained as follows:

Condition (2) shows that not in all regions of an image is there convergence in each point, once the C-product has been applied. If we consider a signal showing some periodicity in a certain region, while in other regions (background) it has a different periodicity, condition (2) allows the separtion of regions in which it behaves differently.

To be precise, in a two-dimentional image A, if we use $\alpha_{i,\,kl}$ to describe an element of the i-th C-set ($C_i$) of image A, then $\alpha_{i,\,kl}$ is represented by a quadruple

$$\alpha_{i,\,kl} \equiv (x_k,\, y_l;\ m_{kl},\, M_{kl})$$

where $x_k$ and $y_l$ are the coordinates of the element (for example, they locate the position of its north-west angle), while $m_{kl}$ and $M_{kl}$ indicate the correspondent max and min values of grey.

If

$$\alpha_{j,\,m\,n} \equiv (x_m,\, y_n;\, m_{mn},\, M_{mn})$$

is an element of the same image, we define the product

$$\beta = \alpha_{i,kl} \otimes \alpha_{j,mn}$$

as $$\beta \equiv (x_p,\ y_q;\ m_{pq},\ M_{pq}) = (x_k\ \ x_m,\ y_1\ \ y_n;\ m_{k1}\ m_{mn},\ M_{k1}\ M_{mn})$$

If the width of the window is w x w, then we can obtain, at most, $w^2$ C-sets and if we perform the C-product $m^2$ times, the final values associated with each pixel are exactly the highest minimum and the lowest maximum in all possible positions of the window containing the pixels.

To visualize the scanning and C-multiplying procedures in two dimensions, we exhibit the process in Fig. 1. To simplify, we choose the width of the window equal to 2 x 2, and use this window to scan the image; we get the sets $C_0$, $C_1$, $C_2$, $C_3$ ......., as described in Fig. 1.1 and we place the couple $(M, m) \equiv (p)$ in the north-west angle of the window, as shown in Fig. 1.2. After the scanning with initial phase (0, 0) a two-dimensional representation of the image by means of the $C_0$-set is obtained, as in Fig. 1.3. We now repeat the scanning of the image with initial phase (0, 1) and get the $C_1$-set. After this scanning we achieve the representation of the image shown in Fig. 1.5, different from the previous one. In the same way, by repeating the scanning with initial phases (0, 1) and (1, 1) we obtain $C_1$ and $C_3$-sets respectively. After repeating the scanning four times, the representation of the image becomes that shown in Fig. 1.8, where each image unit is now filled in by the couple (M, m) belonging to $C_0$, $C_1$, $C_2$ and $C_3$, respectively.

In order to get the best solution, equivalent to the given image, we now introduce the C-product

$$\beta = C_{0,i} \otimes C_{1,j} \otimes \ \ldots\ldots \otimes C_{t,p}$$

as $$\beta \equiv (x,\ y;\ m,\ M)$$

$$= (x_{0,i} \oplus x_{1,j} \oplus \ \ldots\ldots \oplus x_{t,p},\ \ y_{0,i} \otimes y_{1,j} \otimes \ldots ; \otimes y_{t,p};$$

$$m_{0,i} \oplus m_{1,j} \oplus \ldots \oplus m_{t,p},\ \ M_{0,i} \otimes M_{1,j} \otimes \ \ldots\ldots \otimes M_{t,p})$$

where $t = N - 1$; N is the total number of units visible through the window used to scan the image. $C_{0,i}$, $C_{1,j}$, ......, $C_{t,p}$ are the C-set elements overlapping each other.

If the amplitude of the window is equal to 2 x 2, then,

$$\beta = C_{0,i} \otimes C_{1,j} \otimes C_{2,k} \otimes C_{3,1}$$

Fig. 2 describes the C-product of four overlapping elements belonging to the four C-sets. The value of the final C-product is the highest minimum and the lowest maximum level of grey in the area containing the respective C-elements.

Through a window whose dimension is greater than the object, scanning gives elements of the C-set with the same max and min values,so that we are not able to reconstruct the object; while in regions where there are objects bigger than the window, the max and min values differ and the object can therefore be reconstructed.

Consequently, by choosing a convenient dimension for the window we can extract from the image only those parts which we are interested in.

The choice of a suitable dimension of the window, necessary for the extraction of the desired objects from the image, can be determined through a C-transformation [8] or C-matrix [9].

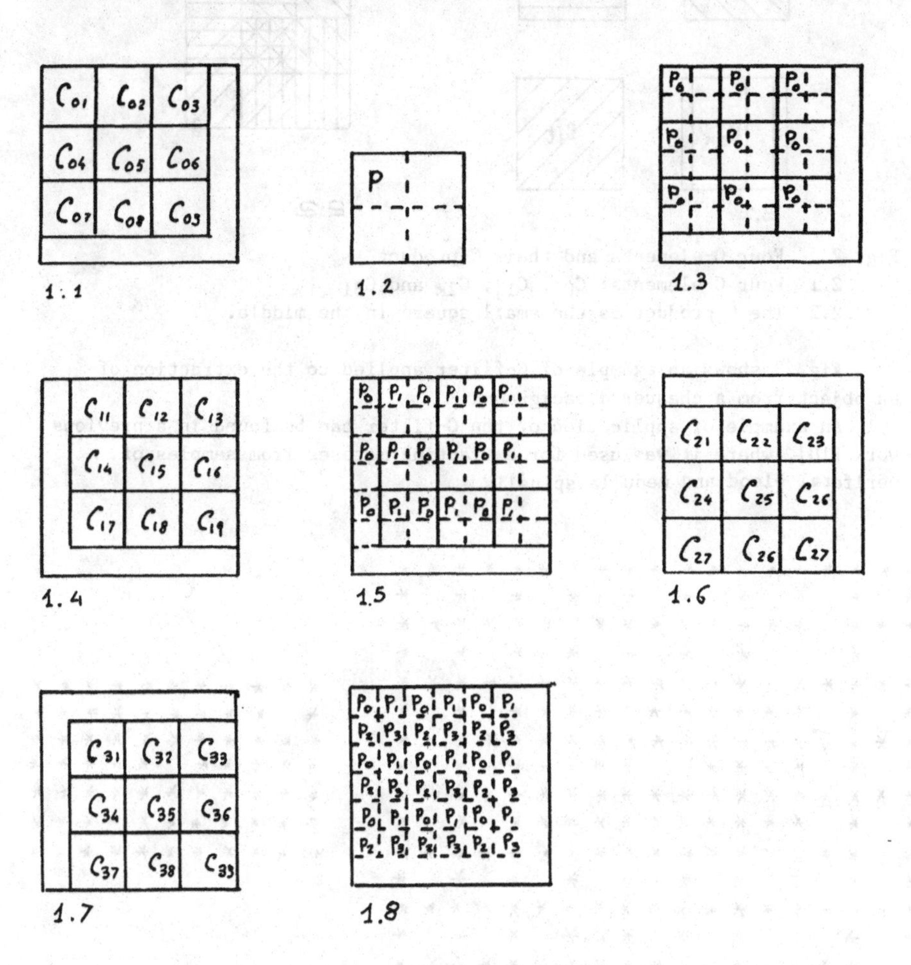

Fig. 1 C-sets and their bidimensional representations

1.1 Elements of $C_0$ with initial phase (0, 0)

1.2 One element of the C-set, the couple M/m (p) is in the north-west angle of the window position

1.3 Bidimensional C-set obtained scanning with initial phase (0, 0)

1.4 Elements of $C_1$ with initial phase (0, 1)

1.5 Bidimensional C-set obtained after the scannings with initial phases (0, 0) and (1, 0)

1.6 Elements of $\overline{C}_2$ with initial phase (0, 1)

1.7 Elements of $C_3$ with initial phase (1, 1)

1.8 Bidimensional C-set obtained after four scannings.

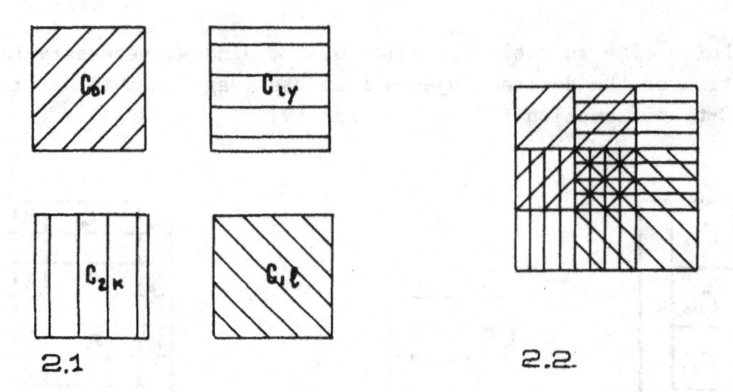


Fig. 2 Four C-elements and their C-product
2.1 Four C-elements: $C_{0i}$, $C_{1j}$, $C_{2k}$ and $C_{3l}$
2.2 The C-product is the small square in the middle.

Fig. 3 shows an example of C-filter applied to the extraction of an object from a chequered background.

An example of application of the C-filter can be found in a previous work [10], where it was used for extracting mitoses from samples of periferal blood and medulla spinalis.

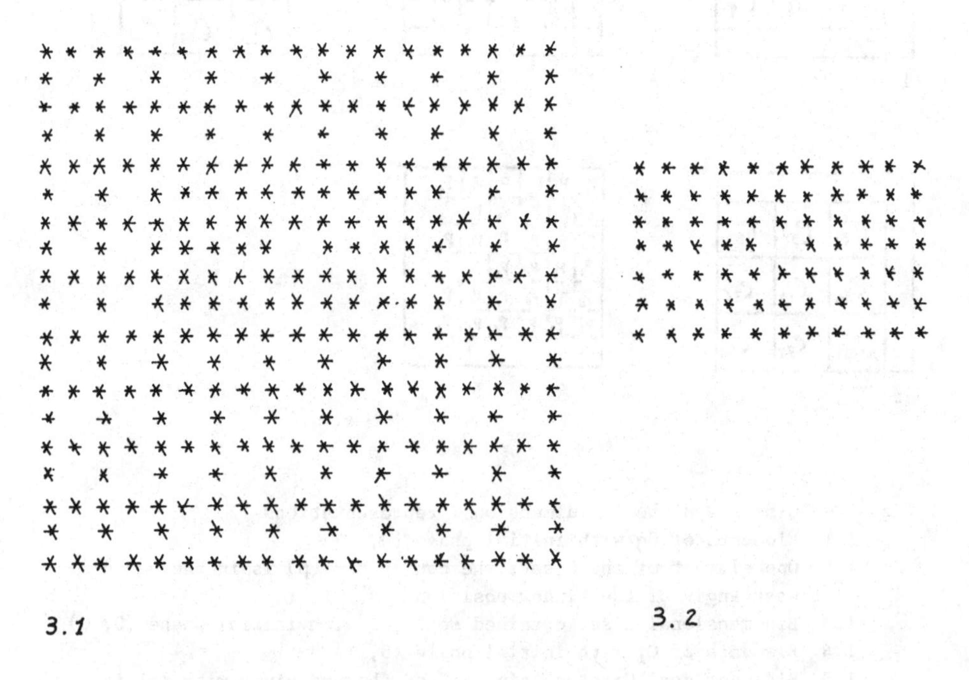


Fig. 3 C-filter
3.1 Before filtering
3.2 After filtering (width of the window 2 x 2)

## C-CONTOUR EXTRACTION

The extraction of the profile or contour of a cell and its nucleus is an important problem for further analysis.

In a binary image, information is completely held in the points on the edges between white and black. The contour extraction method through C-calculus is reported in previous works [7]; there, however, the method presented some inconveniences because it yielded contour lines of the width of two pixels. We introduce here an improved algorithm which yields a contour having the width of only one pixel.

Fix the amplitude of the window as 2 x 2 and perform the scanning of the image in the usual way. At the end, the bidimensional figure (C-set) described in Fig. 8 is obtained. Multiply these C-sets using the standard product rules and apply an algorithm keeping in each product only the elements satisfying the condition

$$\left(\sum_{i=1}^{4} m_i \leqslant 2\right) \wedge (a_{pj} = 1)$$

where $m_i$ is the minimal value of the four overlapping windows and $a_{pj}$ is the pixel in the original image, in the same position as the C-product.

After submitting the C-set elements to this preparation, the contour is obtained. Fig. 4 shows an example of contour.

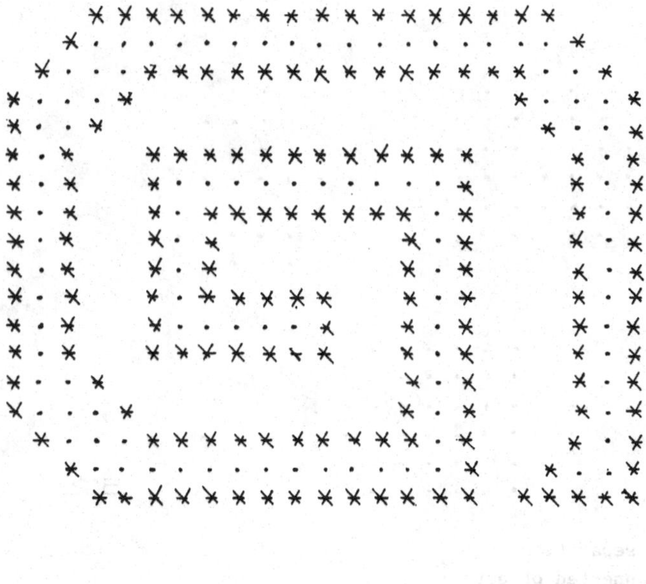

Fig. 4 An object and its contour after contour extraction (represented by the asterisk).

## C-SEPARATION

C-calculus can be used to divide two or more connected objects without altering the objects themselves. This turns out to be useful in the treatment of some microscopic images, when some cells are linked to each other by small bars or by particles deriving from broken cells.

The algorithm is exactly the same as the C-filter. Take the amplitude of the window as W = n x n, with n larger than the thickness of the small bar, but smaller than the diameter of cells. Perform the scanning with this window and take the resulting C-sets; carry out the C-products according to the standard product rules: this will separate the objects. (See Fig. 5)

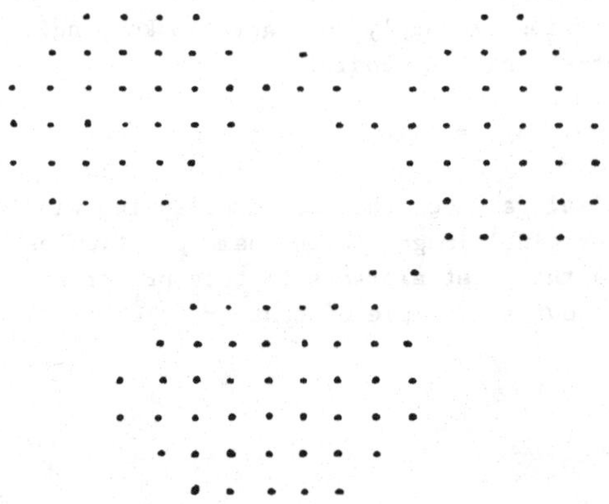

5.1

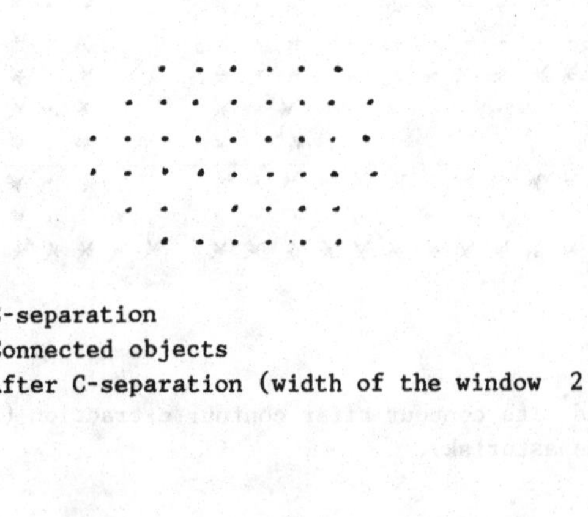

5.2.

Fig. 5 C-separation
5.1 Connected objects
5.2 After C-separation (width of the window 2 x 2)

## C-CONTRACTION AND C-DILATION

Often, especially when microscopic images are tested, we have to deal with digital images of many superposed cells, or cells which touch each other. Examples are granules in a pancreatic cell [11], or superposed nuclei [12].

"C-contraction" can be used to separate such cells or particles. The algorithm for the contraction based on the C-calculus distinguishes the pixels in a rather different manner than that of the usual reduction algorithms; not only edge pixels can be destroyed (usually, reduction algorithms destroy only this part) but also those pixels that have an eight fold connection with zero (background).

We thus define a different C-product, suitable for the C-contraction:

$$\beta \equiv (x_p, y_q; m\ M)$$

$$= (x_k \oplus x_m, y_1 \otimes y_n; \quad m_{k1} \otimes m_{mn}, M = m)$$

Handling of scanning and multiplying operations is the same as standard. The width E of the destryed strip depends on the amplitude of the window W. There is a relationship between E and W:

$$E = W - 1$$

where E is the number of layers destroyed by the C-contraction. Fig. 6 shows the result of using C-contraction.

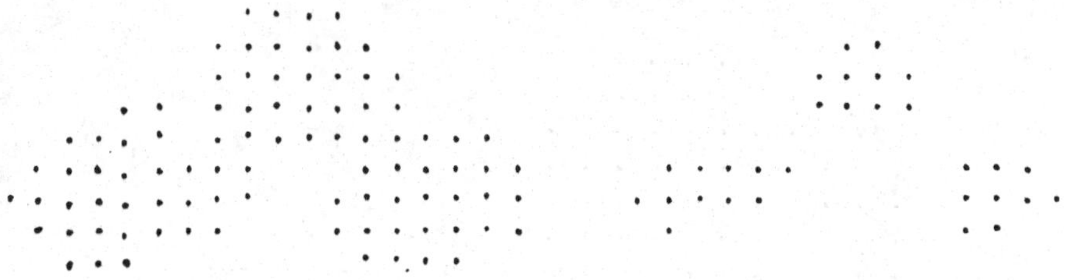

6.1 6.2

Fig. 6 C-contraction
6.1 Touching cells
6.2 After C-contraction (window width 2 x 2)

The function of C-dilation is opposite to that of C-contraction; applying C-dilation, the edge of an object swells towards the background pixels which are connected with the edge pixels.

This function is useful in some cases, when there are holes or cuttings in the image of cells caused by the threshold of the detector and we want to restore the image.

The handling of C-dilation is the same as for C-contraction, except for the C-product which we define in a different way:

$$\beta \equiv (x_p, y_q; m, M)$$
$$= (x_k \oplus x_m, y_l \otimes y_n; m = M, M_{kl} \oplus M_{mn})$$

Fig. 7 shows the result of the application of C-dilation

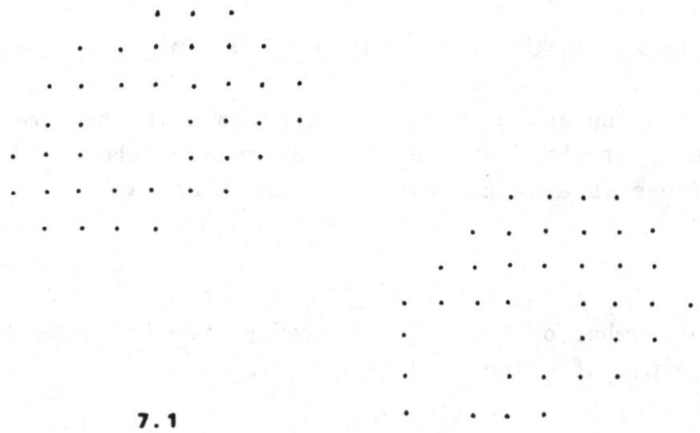

7.1

7.2

Fig. 7 C-dilation
7.1 Before C-dilation
7.2 After C-dilation

## MOVEMENT ANALYSIS

To detect the movement of objects from a sequence of images it is necessary to compare successive photograms. The object seen in a photogram must be searched in the next one; if it is found, and if there is a position change, than it is known to be a moving object, whose speed can be determined by the shifting measured from the two photograms.

Most of the algorithms for the detection of movement are based on a comparison procedure. Some researchers compare unprocessed images by using cross-correlation techniques.

The main limit of these algorithms is their cost for automatic elaboration [14], so that the search for some alternative comparison procedure appears desiderable.

Since in most cases the movement of an object can be sufficiently described by the movement of one of its subsets (the movement of other parts being either redundant or irrelevant), the survey of an object can be concentrated on its "representative regions".

In this work we consider as "representative regions" the contour and the median line of the object.

The extraction of the contour with the C-calculus has been discussed before; we shall discuss next how to obtain the median line of an object with the same technique.

Blum has introduced the symmetrical axis (also called median axis or median line or skeleton) as a particular description of the shape of an object [15]. Several algorithms have been proposed by Rosenfeld [16], Cederberg [17] and others to obtain the skeleton or the median line of an object. Before introducing our algorithm, it is necessary to elaborate some more on C-calculus.

With a given window (of amplitude W x W) the scanning of a binary figure provides the pairs of values of the maximum (M) and minimum (m) levels of grey within the window.

In some areas, where the convergence conditions are satisfied, we obtain by the C-product a unique value for each pixel (i. e., the highest minimum and the lowest maximum coincide), while in areas outside the convergence field we obtain the ambiguous result 1/0.

We choose a window-width as 2 x 2 and scan the median line of an object. Since the median line has the width of one pixel, after scanning around this line we shall obtain the pair M/m with values 1/0 and after multiplication the same ambiguous values 1/0 will remain for each pixel in the position of the median line. The interesting fact is that the analysis of the regions surrounding that where the pairs M/m have values 1/0, distinguishes the parts of an image within one pixel width. Thus, if we use correctly C-contraction to destroy those parts of an image which are thicker than a pixel, the median line of an object can be retrieved.

Our algorithm (C-refinement), in order to obtain the median line of an object, destroys in succession, layer by layer, the pixels of an object which are not critical to retain the connection, until only the median line remains.

The destruction of each layer occurs through the successive elimination of its four sides.

In detail, to carry out a C-refinement, it is necessary:

1. To assign a window (for example, of amplitude 2 x 2) and to scan the binary image in order to get a bidimensional C-set.
2. To test successively each element in the figure, to see if there is a typical cross-shaped pattern (see Fig. 8.1) if so, then it is necessary to control further the central pixel of the cross and its three south-east pixels, and to see which of them vanish. Mark the pixels which are four times connected with null pixels (see examples in Fig. 8.2)

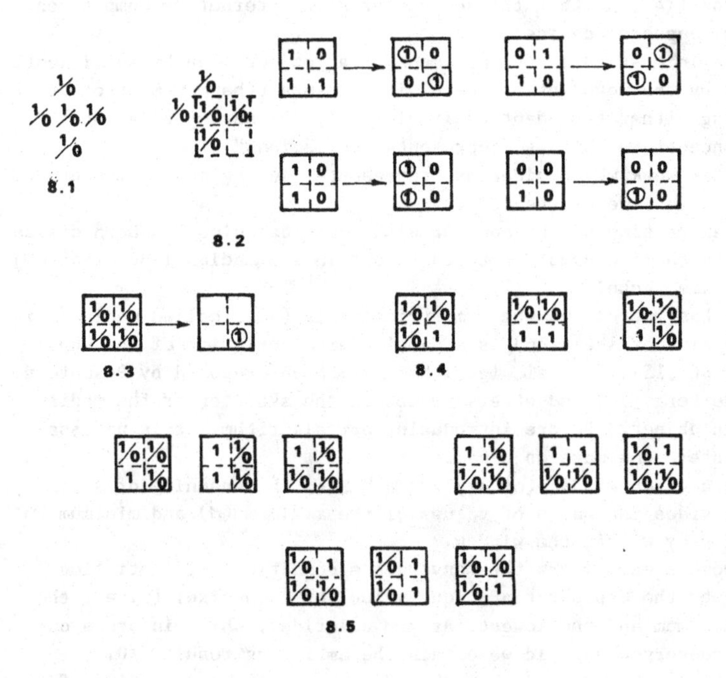

Fig. 8 C-refinement

3. If no cross around an element is found, then check if all the three north-west elements are characterized by the pair 1/0. (see Fig. 8.3). If so, mark the pixel.
4. Through a C-contraction eliminate the north side of the layer. Check in succession each element in the figure; if a scheme similar to that in Fig. 4 results, renew the pixel forming the south-east element of the cross, unless the adjacent pixel in the north is not already marked; in this case, the pixel is to be marked and kept.
5. In order to mark the pixels which are to mantain in critical connection, before destroying each of the other three sides of the layer repeat phases 1 - 3; then, carry out in succession the elimination of the east, south and west sides of the layer, checking every time

the conditions for elimination. Before destroying a pixel it is necessary to check if its east, south or west neighbour, depending on the side which is to be eliminated, is a marked pixel. If so, keep and mark the pixel according to the procedure followed in phase 4.
Then repeat all phases from 1 to 5 until no other pixel is to be destroyed.
An example of C-refinement is shown in Fig. 9.

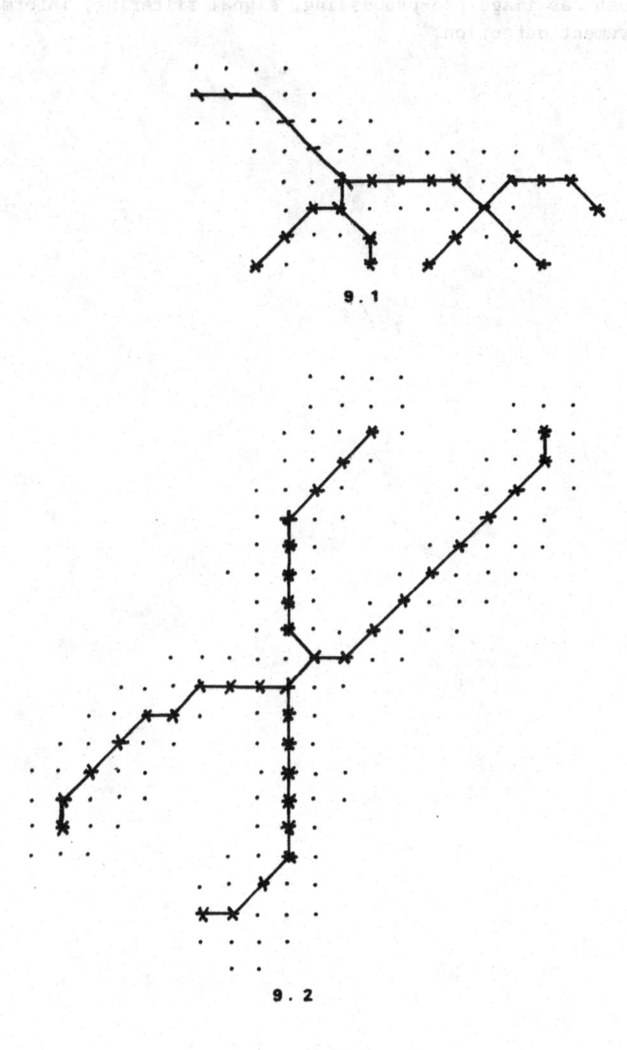


Fig. 9 Examples of C-refinement to obtain a median line of an object (represented by the asterisk)

## CONCLUSIONS

We have various new applications of the C-calculus, such as the C-separation, C-contraction, C-dilation, contour extraction and C-refinement, to the analysis of cell movement, which can be added to the application of the C-filter to the extraction of mitoses from samples [10] and the C-transformation to structure analysis, reported in previous works.

Due to its simplicity and flexibility, the method described here seems to be promising. It does not seem absurd to hope that particularly compact sets of procedures may be obtained from the C-calculus, for use in problems such as image pre-processing, signal filtering, information compaction and movement detection.

## REFERENCES


1. Takagi M. and Sakaue K.: 'The analysis of moving granules in a pancreatic cell by digital moving image processing' - Proc. 4, Int. Jnt. Conf. on Pattern Recognition, Nov. 7 - 10, Kyoto, 735-739 (1978)
2. Ferrie F. P., Levine M. D. and Zucker S. W.: 'Cell tracking and minimization approach' - Proc. 5, Int Jnt. Conf. on Pattern Recognition, Dec. 1 - 4, Florida, 396 - 402 (1980)
3. Pai S. (Bei S.) Sci. Rec., 1, 187 (1942)
4. Bei S. et al.: 'Cell reformation of chirocephalus yolk granules' Scientia Sinica (Series B), 26, n. 5, 454 - 459? (1983)
5. Levine M. D. and Youssef Y. M.: ' An automatic picture processing method for tracking and quantifying the dynamics of blood cell motio' - Department of Electrical Engineering, Mc Gill University, February (1978)
6. Caianiello E. R.: 'A calculus for hierarchical system' - Proc. 1. Int. Jnt. Conf. on Pattern Recognition, Washington, 1 - 2, (1973)
7. Apostolico A., Caianiello E. R., Fischetti E., Vitulano S.: 'C-calculus: an elementary approach to some problems in pattern recognition' - Pattern Recognition, 10, 375 - 387, (1978)
8. Apostolico A., Caianiello E. R. , Fischetti E , Vitulano S.: "An application of C-calculus to texture analysis: C-transform' ibid, , 10, 389 - 395 (1978)
9. Caianiello E. R., Gisolfi A., Vitulano S.: 'C-matrix: a transformation of signals' - Coronado (1979)
10. Caianiello E. R., Gisolfi A., Vitulano S.: 'An identification of mitoses' - Proc. 4, Int. Jnt. Conf. on Pattern Recognition, Nov. 7 - 10, Kyoto, 597, (1978)
11. Sakaue K. and Takagi M.: 'Separation of overlapping particles by iterative method' - Proc. 5, Int. Jnt. Conf. on Pattern Recognition, Dec. 1 - 4, Florida, 522 - 524 (1980)
12. Onoe M., Kuno Y. and Tenjin Y.: 'Recognition of adenocarcinoma in automated uterine cytology' - Proc. 4, Int. Jnt. Conf. on Pattern Recognition, Nov. 7 - 10, Kyoto, 883 - 885 (1978)
13. Smith E. A. and Phillips D. R.: 'Automated cloud tracking using precisely aligned digital ATS pictures' - IEEE Trans. Computers C-21, 715 - 729, (1972)
14. Fennema C.L.: 'Velocity determination in scenes containing several moving objects' - Computer Graphics and Image Processing, 9, 715 - 729, (1979)
15. Blum H.:'A transformation for extracting new descriptors of shape' Models for the Perception of Speach and Visual Form (ed. Walthen Dunn), MIT press, 362 - 380 (1967)
16. Rosenfeld A. and Pfaltz J. L.: 'Sequential operations in digital piture processing' - J. ACM, 13, 471 - 494 (1966)
17. Cederberg R.: ' Shrinking of RC-coded binary paterns' - Proc. 5, Int. Jnt. Conf. on Pattern Recognition, Dec. 1 - 4, Florida, 1019 - 1022 (1980)
18. Caianiello E. R., Gisolfi A. Vitulano S.: 'A technique for texture analysis using C-calculus' - Signal Processing, 1, 159 - 173 (1979)

# SYSTEMS AND UNCERTAINTY: A GEOMETRICAL APPROACH

Eduardo R. Caianiello
Dipartimento di Fisica Teorica e
sue metodologie per le Scienze Applicate
Università di Salerno
84100 Salerno

## 1. DO MEASURABLE QUANTITIES "EXIST"?

Classical mechanics is founded on the assumption that the quantities it studies have an existence of their own, which we may detect and measure, if we care, to any accuracy permitted by experimental know-how. Quantum mechanics sets theoretical limits to the accuracy we may achieve in the measurement of some quantities; their "existence" is not its object, the formalism deals with "phenomena", not "noumena". What about the general systems of Cybernetics (or "General Systems Theory", or any of the more or less synonymic denominations that abound)? A major advance in scientific thought and philosophy has been the realization that all theories are (more or less successful) "models": man-made, that is, so that the metaphysical notion of "truth", attached in turn to Newton's or Einstein's descriptions of gravitation, to mention the most classic example, has lost its absolute connotation. The search for better, or just alternative models for gravitation and everything else fills all scientific journals.

Whenever measurements of data related to some model are performed, "errors" or "noise" are of course expected, as unavoidable limits to the attainable precision. Quantum mechanics's "uncertainty" is a limitation of a different sort: it has been embedded in the model. Is this a special case? Can we assume, say, that the "value of the dollar" at the Zurich exchange at a precise time is something one could measure, if one wished to do so, with arbitrary accuracy with, for instance, four hundred exact digits? The cost of the apparatus one should set up (if conceivable) to perform such an experiment would, it can hardly be doubted, alter that same value, not unappreciably. Does then such a thing as the "value of the dollar" have an existence of its own, or is it model-dependent, as is the case with the formalism of quantum mechanics? R.E.Kalman [1] has proved that even in the simplest case of a two-dimensional linear model one obtains quite different results in the hypotheses: a) of arbitrary measurability for both parameters, b) of model dependence. He also states that this fact is totally ignored in extant system theory. "Uncertainty" cannot be swept under the carpet



*E. R. Caianiello and M. A. Aizerman (eds.), Topics in the General Theory of Structures, 199–206.*
*© 1987 by D. Reidel Publishing Company.*

in any model, econometric or of any other sort, without falling into the fallacy that quantum mechanics has avoided at last. There has been an almost astounding coincidence between his remarks, and the results obtained by this writer and his co-workers [2] in attempts to obtain a geometrical representation of quantum mechanics, in which uncertainty appears as "curvature" in (relativistic) phase space. E.T.Jaynes's principle of "maximal entropy" [3] (modified to cross-entropy, so that it becomes synonymous with "information") together with the well-known geometrization of information theory (to be quoted later) are the keys to this approach: a complexified information theory can be proved, in fact, to extend the results obtained by present "pre-quantum geometry" [2,4], and to yield, besides standard quantum mechanics, additional previsions (such as mass spectra for particles, a maximal acceleration that, combined with the equivalence principle, yields directly A.D.Sacharov's maximal temperature, etc.).

We wish to emphasize here that, for our present purposes, the case of quantum mechanics is considered only because it certainly constitutes the most exacting test to which we could subject our thesis. In the context of Cybernetics, it is to be considered only as an example, hopefully a paradigm for the study of many more diverse situations. It has of course the advantage that "uncertainty" is fixed in it in a simple and neat way by Heisenberg's relations.

## 2. FROM ENTROPY TO CROSS-ENTROPY

There are at least two things wrong with Shannon entropy. The first is its name: had Claude Shannon not accepted John von Neumann's advice ("you should call it entropy and for two reasons: first, the function is already in use in thermodynamics under that name; second and more importantly, most people don't know what entropy really is, and if you use the word "entropy", in an argument, you will win every time" [5]) and just called it "uncertainty", endless confusion would have been spared. "Entropy" is psycologically tied with "thermodynamics" in a physicist's mind, so that the purely logical, far wider connotation of Shannon's concept escapes attention. Shannon entropy is simply and avowedly the "measure of the uncertainty inherent in a pre-assigned probability scheme"; as such it has nothing whatever to do with thermodynamical entropy, except in the case in which the probability distribution is known, or proven to be, "canonical".

Shannon entropy goes wrong in a second respect. Whenever one deals with a continuous probability distribution $\rho(x) \geqslant 0$

$$(1) \qquad \int_{-\infty}^{\infty} \rho(x)\,dx = 1$$

$$(2) \qquad H = - \int_{-\infty}^{\infty} \rho(x) \lg \rho(x)\,dx$$

Shannon entropy diverges (it is also not invariant under change of variables $\rho(x) \rightarrow \rho(x(y))\frac{dx}{dy}$). It can, though, be changed equivalently into

$$(3) \qquad H_c = \int \rho(x) \lg \frac{\rho(x)}{\rho_0(x)} dx$$

(we change the sign here to conform to usage), the so-called "cross-entropy". It is exempt from all the troubles mentioned before, especially if we decide to call it (with one of the many uses of the term) "information". Contrary to Shannon entropy, which gives the amount of uncertainty connected with a single probability distribution (i.e. the average "self-information - lgo", an essentially static concept), the cross-entropy (3) measures the uncertainty connected to a posterior distribution $\rho(x)$, once a prior distribution $\rho_0(x)$ is known (ample room for typical statistical inference, and dynamics).

The classic works of Jaynes [3] and others maximize (2) or (3) subject to conditions chosen so as to obtain canonical (Boltzmann, Bose, Fermi, Gentile) statistics.

## 3. JEFFREY'S DIVERGENCE AND INFORMATION METRIC

Information, entropy, cross-entropy etc. have been defined in a great many ways, most of them known only to specialists. Of particular interest to us here is the case of parametric distributions $\rho(x|z)$. Here $z \equiv \{z_{(1)}, z_{(2)}, \ldots, z_{(m)}\}$ belongs to an $R^n$ random space and $x \equiv \{x^{(1)}, x^{(2)}, \ldots x^{(m)}\}$ to an $R^m$ parameter space. For example, with the gaussian

$$(4) \qquad \rho(x|z) = \frac{1}{\sqrt{2\pi}\sigma} \exp\left[-\frac{(z_1-\mu)^2}{2\sigma^2}\right]$$

we read

$$z \equiv \{z_{(1)}\} \quad ; \quad x \equiv \{x^{(1)} = f^{(1)}(\mu,\sigma),\ x^{(2)} = f^{(2)}(\mu,\sigma)\}$$

The cross-entropy (3) is called in this context "Kullback-Leibler information" and is written

$$(5) \qquad H_c = I(1,2) = \int dz\, \rho(x_1|z) \lg \frac{\rho(x_1|z)}{\rho(x_2|z)}$$

Expression (5) measures a "distance" between distribution $\rho(x_2|z)$ and $\rho(x_1|z)$ (e.g. between different gaussians). Owing to its asymmetry it is called the directed distance $2 \rightarrow 1$. Its symmetrized form

(6) $J(1,2) = I(1,2) + I(2,1)$

is called the J-divergence (after Jeffreys [6]) or "distance"between $\rho(x_2|z)$ and $\rho(x_1|z)$.
As numbers, (5) and (6) evaluate what their names indicate. The term "distance" however is not wholly appropriate. The triangle inequality applies only in the infinitesimal case, in which $(x_1 = x;\ x_2 = x + dx)$

(7) $J(x,x+dx) = ds^2 = g_{hk}(x)dx^h\ dx^k$

Here, with the notation $\partial_h = \frac{\partial}{\partial x^h}$, we have

(8) $g_{hk}(x) = g_{kh}(x) = \int dz\ \rho(x|z)\,\partial_h \lg \rho(x|z)\,\partial_k \lg \rho(x|z)$

the well-known information (or Fisher) metric [7]. Of course $\rho(x|z) \geqslant 0$ and

(9) $\int \rho(x|z)dz = 1$

The concepts of Fisher metric, information and infinitesimal distance have led to many classic developments in information and estimation theories, with Riemannian geometry the "natural tool". "Cross-entropy", or "information distance", through (7) and (8) provides a natural metrization of otherwise affine space elements, prior to any introduction of physical concepts such as "energy", "temperature" etc. The work of Amari [8] and others shows, in fields other than physics, the fruitfullness of utilizing in statistical inference the logical principle of minimizing cross-entropy, i.e. choosing the least biased distribution permitted by the circumstances. It has also been called the "Principle of Maximum Honesty" [9].

The ingredients in all work of this kind are, whatever the ideology behind it:

1) extremizing entropy, or cross-entropy, information;

2) doing so with appropriate constraints (Lagrange multipliers etc.)

Step no.1 carries on, formally, regardless of whether the domain over which the integration is extended is the whole, or a small "infinitesimal" region of it. Step no.2 requires instead that integrals be taken on the whole domain.

The end products are the well-known distributions of mathematical statistics, thermodynamics etc. A typical estimation problem is, for instance, the determination of the numerical value of a parameter which specifies, within a given parametric family, the distribution that best fits some wanted requirements. (Canonical distributions have thermodynamic interpretations.)
There is nothing, however, against adopting the same procedure with

a different outlook, which we wish to emphasize here. One gives up, at the beginning, step no.2. Step no.1, alone, gives a metric G; a geometrical model can then be set up in the standard way by assigning a connection $\Gamma^{\alpha}_{\mu\beta}$ (not in general symmetric in $\mu$ and $\beta$ ), from which curvature is defined. The demands that previously were handled with step no.2 can now be met by a correct choice of G , $\Gamma$ and whatever else geometry requires. This approach is standard in geometrical work on estimation [8].

At this stage, however, we can also look at this model as a standard source of wave equations, geodesics, and all that. We have shown[2] that there is a general geometric framework, best expressed in complex spaces, in which the natural objects to compute are "square roots of probabilities", i.e. wave functions. both the classic geometry of information theory mentioned before and the "quantum geometry" developed by us in phase space along these lines [4] appear as subcases.

The information metric (8) can be identically written

$$(10) \qquad g_{hk}(x) = 4\int dz\, \partial_h \sqrt{\rho(x|z)} \cdot \partial_k \sqrt{\rho(x|z)}$$

the most general connection compatible with information geometry is [10]

$$(11) \qquad \overset{\alpha}{\Gamma}_{ijk} = [ijk] - \frac{\alpha}{2}\int dz\, \partial_i \lg\rho \cdot \partial_j \lg\rho \cdot \partial_k \lg\rho$$

$\alpha$ arbitrary real. One has then, e.g. for the Gaussian distribution (8)

$$(12) \qquad G = \begin{pmatrix} \sigma^2 & -2\mu\sigma^2 \\ -2\mu\sigma^2 & 4\mu^2\sigma^4 + 2\sigma^4 \end{pmatrix}$$

and

$$(13) \qquad \overset{\alpha}{R}_{1212} = (1-\alpha^2)\sigma^6$$

Of interest to us is the case $\alpha = 0$, the only one that guarantees the compatibility, or metricity of the connection with G; i.e. vanishing co-variant derivatives of G.
Then one finds that

$$(14) \qquad \overset{\circ}{\Gamma}_{ijk} = 4\int dz\, \partial_i \partial_j \sqrt{\rho}\; \partial_k \sqrt{\rho}$$

$$(15) \qquad \overset{\circ}{R}_{1212} = \sigma^6$$

the curvature tensor expresses our uncertainty, as it vanishes only when $\alpha = 0$. We see therefore that a Riemann tensor offers a natural way of expressing uncertainty.

Quantum mechanics appears as a special case, privileged by the fact that Heisenberg's uncertainty relations are very neat and specific. We have used them indeed to define a curvature tensor in phase space (enlarged to t and E ), from which all standard results of quantum mechanics are verified to derive, with several additions and generalizations. Such a scheme, wide enough to accomodate both information and a quantum geometry, is suggested by the classic form (10). With some hindsight we generalize it to an hermitian metric, e.g. like

$$(16) \qquad g_{hk}(x) = \bar{g}_{kh}(x) = \int dz\, \Psi_h(x|z)\, \overline{\Psi}_k(x|z) \qquad ;$$

if z is discrete, (14) becomes

$$(17) \qquad g_{hk}(x) = \sum_{\lambda} \Psi_{(\lambda)h}(x)\, \overline{\Psi_{(\lambda)k}(x)}$$

a viel-bein, which becomes real and holomonic if

$$\Psi_{(\lambda)h} = \overline{\Psi_{(\lambda)h}} = \partial_h \varphi_{(\lambda)}(x)$$

and clearly $\left( \int dz \equiv \sum_{\lambda} \right)$ reduces then to the information metric.

## 4. THE HEISENBERG AND CRAMER-RAO INEQUALITIES

Mention is made here of yet another connection of general systems with geometry. The information metric tensor just described satisfies the celebrated Cramér-Rao [11] inequalities. It suffices to recall here the one-dimensional case; going from the sophisticated language of information theory to that of physics, i.e. writing, if $\tilde{x}$ is an estimator, $\mathrm{var}_x(\tilde{x}) = (\Delta x)^2$, they reduce to ($g_{11}$ is also called "Fisher information" [12]).

$$(18) \qquad (\Delta x)^2 . g_{11} \geqslant 1$$

If $\rho(x|z) = \rho(x-z)$, etc., the form (10) of $g_{11}$ yields

$$(19) \qquad (\Delta x)^4 . \; 4 \int dz \left( \frac{\partial \sqrt{\rho}}{\partial x} \right)^2 \geqslant 1$$

i.e. with $\sqrt{\rho} = \varphi \quad , \quad p_x = -i\hbar \frac{\partial}{\partial x}$

$$(20) \qquad \Delta x . \Delta p_x \geqslant \frac{\hbar}{2}$$

Uncertainty relations appear as special cases of the Cramér-Rao inequalities, not invented purely for physics. Text-book quantum mechanics could be obtained, as it is, from these inequalities. Our model, in which curvature in phase space expresses uncertainty, proves to be more general (in particular, it introduces naturally non-Abelian as well as Abelian gauge fields [13]).

## 5. THE SIGN OF $ds^2$

To conclude, consider ordinary Euclidean distance. A theory predicts average positions (or whatever): actual measurements give a scatter around them. Suppose that repeated measurements, according to book, yield gaussian distributions of observed $z_x$, $z_y$, $z_z$ values around the corresponding *average* position ($\mu_x \equiv x, \mu_y \equiv y, \mu_z \equiv z$ : our parameter space): dispersion (14) is supposed to be the same in all directions; then (4) now reads

$$(21)\quad \rho(x,y,z \mid z_x,z_y,z_z) = \frac{1}{(\sqrt{2\pi}\sigma)^3}\exp\left[-\frac{1}{2\sigma^2}\left\{(z_x-x)^2+(z_y-y)^2+(z_z-z)^2\right\}\right]$$

we find from (10)

$$(22)\quad ds^2 \equiv dH_c = \left|\text{const}\right|^2.(dx^2+dy^2+dz^2) \geqslant 0$$

In special relativity we would find

$$(23)\quad ds^2 \equiv dH_c = \left|\text{const}\right|^2.\left( dt^2-\frac{1}{c^2}\left[dx^2+dy^2+dz^2\right]\right) \geqslant 0$$

and in our quantum geometry [2]

$$(24)\quad ds^2 \equiv dH_c = \left|\text{const}\right|^2.\left\{dt^2 - \frac{1}{c^2}d\vec{x}^2 + \frac{\hbar^2}{\mu^4 c^6}\left[\frac{dE^2}{c^2} - d\vec{p}^2\right]\right\} \geqslant 0$$

It appears that $dH_c \geqslant 0$ , which is a *consequence of the mathematical expression of the cross-entropy*, not only imposes the known requirement for *phenomena* to be physical, but, read as $ds^2$, also for *distances* to describe physical particles. (22) and (20) need no comment; (24) is the condition that leads, in our quantum geometry, to the introduction of a *maximal acceleration* [14], which, if use is made of the equivalence principle, leads directly to A.D.Sacharov's maximal temperature [15], without additional assumptions, as a quantum limitation [16]. The present approach leads (c.f. our works cited in the references) to a "quantum geometry" in which "$ds^2$", and hence lagrangians etc., is an "information distance": all mathematical tools introduced derive from Jaynes's principle (as here re-interpreted), which belongs to *"logic"*, not to "dynamics" (as it was not "thermodynamical" before). We propose, that is, a *universal logical paradigm* that, through extremization of information or cross-entropy, should be applicable in principle to all sciences studying "phenomena" (rather than "noumena" as does classical mechanics). That the supplementary information needed to construct a model be given in the form of overall constraints, or by assignment of appropriate metrics and connections, becomes here a technical, not a fundamental issue.

The novel conceptual basis appears here to be that if I know "something" about a quantity, that is *all* I know. The measurement of a quantity *creates* that quantity and the value that uncertainty permits for it.

In other words, notions such as "space", or "value of the dollar here and now", are, alike, *model-dependent*: they have no existence or meaning of their own before measurement, and different measurements may generate quite different notions and values. *Measurement creates the thing it measures.*

References

1. R.E.Kalman: Proc.Int.Symp. on Dynamical Systems, ed. A.Bednarek: Current Developments in the Interface: Economics, Econometrics, Mathematics", ed. M. Hazewinkel and A.H.G. Rinnooy Kan (Dordrecht, D. Reidel, 1982).
2. E.R.Caianiello: Lettere Nuovo Cimento, 25, 225 (1979); 27, 89 (1980); 38, 539 (1983); 35, 381 (1982); Nuovo Cimento, 59B, 350 (1980); Proc.IV Conf. on Quantum Theory and the Structure of Space and Time, Tutzing (1981); Proc.VI JINR Int. Conf. on Problems of Quantum Theory, Alushta (1980); (with G.Vilasi); Lettere Nuovo Cimento, 30, 469 (1981); (with G.Vilasi and S.De Filippo) ib, 33, 555 (1982); (with G.Marmo and G.Scarpetta), ib, 36, 487 (1983); several other papers in print.
3. E.T.Jaynes: Phys.Revs., 106, 620 (1952); in Brandeis Theor.Phys. Lecture on Statistical Physics, 3 (New York).
4. E.R.Caianiello: Lettere Nuovo Cimento, 38, 539 (1983).
5. M.Tribus: Rational Descriptions decisions and designs, Pergamon Press, Oxford, (1969).
6. H.Jeffrey: Theory of Probability, 2nd ed., Clarenton Press, Oxford, (1948).
7. R.A.Fisher: Phil.Trans.Roy.soc., 22A, 309 (1921); Proc.Cambridge Phil.Soc., 122, 700 (1925).
8. S.Amari: Raag.Reps., 106, Feb.1980; Techn.Reps.Fac.Eng.Univ.of Tokyo, METR 81-1, April 1981, ib METR 84-1, Jan.1984;
B.Efron: Ann.Statist., 3, 1189 (1975); 6, 362 (1978).
A.P.David: Ann.Statist., 3, 1231 (1975); 5, 1249 (1977).
9. J.N.Kapur: Journ.Math.Phys.Sci., 17, 103 (1983).
10. N.N.Chentzov: Statistical Decision Rules and Optimal Conclusions (in Russian) Moscow, (1972).
11. C.R.Rao: Linear statistical Inference and its applications, J.Wiley, New York, (1973) and papers quoted therin.
12. R.A.Fisher: cf. Ref.7,b.
13. E.R.Caianiello, G.Marmo, G.Scarpetta: '(Pre)quantum Geometry', Nuovo Cimento, 38, 539 (1983).
14. E.R.Caianiello: Lettere Nuovo Cimento, 32, 65 (1981); (with S.De Filippo, G.Marmo, G.Vilasi), ib, 34, 112 (1982); others in print.
15. A.D.Sacharov: JETP Letters, 3, 288 (1966).
16. E.R.Caianiello, G.Land: Lettere Nuovo Cimento, 42, 70 (1985).

SUBJECT INDEX